# MS.
# ADVENTURE

# MS. ADVENTURE

## MY WILD EXPLORATIONS IN SCIENCE, LAVA, AND LIFE

## JESS PHOENIX

TIMBER PRESS

PORTLAND, OREGON

Published in 2021 by Timber Press, Inc.
The Haseltine Building
133 S.W. Second Avenue, Suite 450
Portland, Oregon 97204-3527
timberpress.com

Printed in the United States of America
Text design by Vincent James
Jacket design by Amanda Weiss
Jacket cover photograph by Adobe Stock/willyam

ISBN 978-1-64326-003-7

Catalog records for this book are available from the
Library of Congress and the British Library.

To Carlos, who always
welcomes our adventures.

This work is my love letter to science,
to all it has given me, and to all it
can give to every single person on
our marvelous, complicated, by turns
knowable and inscrutable planet.
We are all born as scientists.
Cell phones, house cats, and Stephen
Hawking—we are all made of the
same stardust. It is what we do
with it that counts.

# Contents

# 1

## Welcome to the Club

MARCH 2016

"THERE IT IS!"

The words tumbled out as I pointed to a building flying a unique flag on a black pole jutting from its brick face. A patch of bright red formed a right triangle on the side closest to the pole, while the middle section contained a white rhombus with three red nautical-looking symbols for contrast. The section farthest from the pole was another right triangle, this one flipped upside down and deep blue, the color of the ocean after a storm.

The Explorers Club.

My heart thumped, and I felt my face and neck flush—a side effect of pale skin. Glancing down, I smoothed my dress into place, cursing my choice of such uncomfortable, feminine clothing. The dress was long-sleeved, jersey knit, black with thin white horizontal stripes. It had a scoop neck and hit just above the knee. It was a perfectly fine dress for a slightly fancy afternoon cocktail social event, but I felt like a gorilla wearing a tutu and heels trying to attend a wedding with the Queen of England.

For the fifth time, I asked Carlos if my outfit was acceptable. He smiled and said I looked great. I sighed, resigning myself to his inherent spousal bias and continued crossing Madison Avenue. I was on high alert, scanning the faces of approaching walkers for clues that they were part of the legendary fraternity, the Explorers Club, that awaited. The most likely members would be older men with worn, creased faces and the occasional missing digit, the stories of challenges met and extremes mastered written on their skin—so I surmised.

I tasted grime as we made our way through the throng of New Yorkers yammering into phones about Friday night plans and dodging pungent piles of garbage. On this day, none of the chaos of the city registered. I was in another world, mentally. Spring in New York—the city of endless possibility, the core of so many stories of optimism that undergird its appeal to so many Americans, new and old—has always felt like a waking dream to me. It's a promise never quite fulfilled like it is in more natural landscapes. The abundant concrete restricts the capacity to bloom. Today, however, even the claustrophobia of the impassive brownstones standing sentinel over East 70th Street failed to temper my excitement. I was in New York not as a tourist or a supplicant, seeking to gain something from the Big Apple. I was there because, for the first time in my life, I was a part of the heartbeat of that carefully constructed monument to humankind's dreams.

Carlos asked if I wanted a picture in front of the building, and I agreed, trying to normalize the situation in my mind. I pasted on a smile while he lined up the shot, a dozen awful scenarios flashing behind my eyes. What if they had no record of me? What if I was dressed too casually? Or overdressed? What if the first person I saw was Buzz Aldrin? What if there had been a mistake and my scientific work actually *hadn't* been deemed good enough for membership? What if no one wanted to talk to me? What if, what if, what if. The drumbeat of doubt threatened to turn my feet toward Central Park, the largest approximation of nature remaining in the concrete jungle that had swallowed the island of Manhattan.

"One, two, three!"

Carlos snapped the picture and handed me the phone for inspection. My face looked ghostly against the black dress; the curse of the redhead made worse by my lackluster makeup skills. I turned toward the club and Carlos squeezed my hand. He asked if I was ready. My eyes widened as I plunged forward, the arched steel doors a final barrier between me and the giants of exploration.

▲

The Explorers Club was founded in 1904 as the United States'
answer to the prestigious Royal Geographical Society of the
United Kingdom. The club's members have laid claim to a lit-
any of famous firsts—first to the North Pole, first to the South
Pole, first solo flight across the Atlantic Ocean, first to the summit of Mount Everest, first to the deepest part of the formidable
Mariana Trench, and yes, first to the surface of the Moon. The
club's founders were active in exploration work around the world
and created the organization to support and promote exploration in any way possible. Returning explorers gave both public
and members-only lectures about their experiences in far-flung
lands, a tradition that continues today.

The specter of colonialism hung over the club for decades,
and the "trophies" present in the club's collection are a vivid—
and at times unsettling—testament to the historical practice of
collecting rare animals and unique cultural artifacts. Like many
institutions that have withstood the painful reckonings of the
twentieth century, the Explorers Club had to shift both its mission and its membership criteria to remain relevant. In 1981,
famed scientist Carl Sagan authored a feminist letter exhorting
the club to admit women to its ranks:

> When our organization was formed in 1905, men were
> preventing women from voting and from pursuing many
> occupations for which they are clearly suited. In the popular
> mind, exploration was not what women did. Even so, women

had played a significant but unheralded role in the history of exploration. . . .

Today women are making extraordinary contributions in areas of fundamental interest to our organization. There are several women astronauts. The earliest footprints—3.6 million years old—made by a member of the human family have been found in a volcanic ash flow in Tanzania by Mary Leakey. Trailblazing studies of the behavior of primates in the wild have been performed by dozens of young women, each spending years with a different primate species. Jane Goodall's studies of the chimpanzee are the best known of these investigations which illuminate human origins. The undersea depth record is held by Sylvia Earle. The solar wind was first measured *in situ* by Marcia Neugebauer, using the Mariner 2 spacecraft. The first active volcanos [sic] beyond the Earth were discovered on the Jovian moon Io by Linda Morabito, using the Voyager 1 spacecraft. These examples of modern exploration and discovery could be multiplied a hundredfold. They are of true historical significance. If membership in The Explorers Club is restricted to men, the loss will be ours; we will only be depriving ourselves.

His effort worked. Sagan's letter also reflects the shift in the club's mission. Rather than simply traversing new or difficult places, which often already had a history of habitation by indigenous peoples, the club moved to promoting exploration that would add to the body of scientific knowledge about our world and our place in it. It became a club that was not simply a social group for wealthy white men, but a place designed to foster the spirit of exploration and essential scientific discovery for generations to come. Among its members and honorees are dozens of household names: Sir Edmund Hillary, Tenzing Norgay, Roald Amundsen, Jacques Piccard, Neil Armstrong, Buzz Aldrin, Sylvia Earle, Sally Ride, Jane Goodall, Theodore Roosevelt Jr., James Cameron, Jim Lovell, and more. Nowadays, the club seemed to strike a balance between the stolid science of academia and the epic feats of the adventure stories I'd grown up on. If Robinson Crusoe, Captain Nemo, Indiana Jones, and James T. Kirk had flesh-and-blood counterparts, they would be in the Explorers Club.

With membership globally hovering just above three thousand, it is an insular group with a secretive air due to the caliber of its famous members and the restrictions it places on visiting members of the public. Only the first two floors of the six-story New York City headquarters are typically open to nonmembers, and the artifacts visible on those floors are enough to wow even the most cosmopolitan visitors. A stuffed Siberian polar bear, standing upright and perpetually snarling, the globe around which Thor Heyerdahl planned the legendary Kon-Tiki raft expedition across the Pacific Ocean, a table made from the hatch of the survey ship *Explorer*, which survived the Pearl Harbor attack due to a timely absence, an ornate carved wooden chair that belonged to the wife of China's last emperor, ice axes hanging over the members-only bar where countless tales of explorations past have been related over evening beverages— even the décor is legendary.

To become part of the Explorers Club, you must first identify the level of membership you qualify for: fellow, member, associate, or student. Every level requires sponsorship from a current club member. Each has its own distinct requirements, which are intentionally stringent to ensure members are truly exploring in the name of science, rather than partaking in adventure travel, big-game hunting, or art photography. Tourist safaris do not qualify for membership, regardless of how extreme they may appear.

I knew I had worked in remote and extreme places, and all my travels (save one misjudgment) had been for science. Acutely aware of the public relations problem scientists have had over the decades, I hoped belonging to the club would help me reach a new audience to communicate the importance of science. Work on natural hazards and climate change was gaining urgency and garnering new public support—a little romanticizing of scientific fieldwork could only help advance our pleas for funding and attention. Still, I was less than half the age of most members, and working on volcanoes, glaciers, and in the shadow of human danger was par for the course in geology circles. I had never punched a Nazi or amputated my own appendages. I had not been to the Moon, discovered a new species of animal, or won

a Nobel Prize. Would I be qualified for the one club I so desperately wanted to join?

▲

The foyer rippled with activity. Gray-clad workers pushed
wheeled carts up a temporary ramp that crested a few low
stone stairs 20 feet past the entry. Beyond the ramp, a massive
globe topped with a heavy-duty protective plastic cover rested
in a squat stand with solid wooden legs. Everything permanent
seemed to have ornamentation—even the ceiling had decora-
tive plaster molding. A large arched doorway gaped open, the
room inside dimmer than the foyer. Affixed above the opening was a brass plaque that read "Club Lounge and Library for
Members Only." To the right of the door was a smaller brass
plaque inscribed with the words "Members Lounge." Below it,
a wrought-iron floor stand held a sign with strong white letters
declaring "Members Only," obviously an afterthought to keep
curious visitors in public-friendly areas.

I peered into the wood-paneled room beyond the signs, and I
could just make out a 5-foot-tall elephant tusk standing upright
next to a fireplace. An ornate wall sconce holding an electric light
styled after a large candle inhabited the space above the tusk. It
emitted soft amber light, lending the room a bygone atmosphere.

Back in the foyer, a behemoth dark wooden desk with the
club seal on its front dominated the right side of the room. It sat

a few feet away from the wall, which had another fireplace that was obstructed by a couple of rolling office chairs. A woman in her mid-twenties with tawny hair and stylish glasses sat in one of the chairs, a phone pressed between her ear and her left shoulder while she typed on a computer tucked behind the tall desk. She glanced up, meeting my eyes while finishing her conversation. Replacing the phone on its base, she typed a few more words and then looked up at us brightly.

"How can I help you?"

She sounded genuine, which eased my tension. I told her I was a new fellow in the club, and that Carlos was my husband and plus-one for the evening. She welcomed me and started to return to her work, but my uncertainty must have been evident. Looking up again, she told us to go ahead and look around.

"Really?" I was incredulous at the informality.

"Of course! It's your club."

Damn. I was *in*.

▲

When I applied to the club in the summer of 2015, I decided to be bold and go for fellow status: the highest level of membership and the one with the fewest number of members. To achieve this level, the applicant must demonstrate contributions to the published body of scientific knowledge via exploration or fieldwork and obtain two letters of sponsorship from club members or fellows.

I didn't know any club members personally, so as happens
so often in the first several years of a career, I needed to ask for
help from someone who had lent a hand before. On LinkedIn I
found a second-degree connection to a fellow named Bill Steele
through Jack Hess, a geologist whose work I respected tremen-
dously. I had made Jack's acquaintance in late 2013, just after
Carlos and I cofounded the nonprofit environmental science
research and education organization Blueprint Earth. Jack was
the executive director of the Geological Society of America at the
time, and he helped Blueprint Earth's work gain early legitimacy.
Jack obliged with the contact, and I left a voicemail for Bill. After
a few days with no response, the club faded to the back of my
mind. I thought it was a long shot, so I was not surprised with
the lack of response.

Several weeks later, I was pulling out of a parking lot when
my phone rang. I answered, and the deep voice on the line introduced himself as Bill Steele. He apologized for not returning my
call sooner and offered the best excuse possible—he had been
underground for the last few weeks in the Western Hemisphere's
deepest cave, Sistema Huautla in Oaxaca, Mexico. I immediately steered the car onto the shoulder to better concentrate on
the call.

Bill told me a bit about his exploration work and his
decades-long career working for the Boy Scouts of America
before asking about my scientific work and why I wanted to join
the club. I have no recollection of exactly what I told him, but

I must have mentioned my work on active volcanoes and with Blueprint Earth. To my delight and no small degree of surprise, Bill offered to sponsor me and requested that I email him my curriculum vitae. I asked if he knew anyone who would agree to be my second sponsor, and he said that he was certain his friend Jim Smith, a geologist and fellow caver, would be happy to.

I hung up the phone in a state of elated shock. Impostor syndrome is very real, and particularly so because I made the uncommon leap from humanities to the hard sciences.

▲

I entered Smith College in the fall of 2000 determined to be an English professor, specializing in the work of modernist poet and writer T. S. Eliot. An early run-in with a member of the English department was severe enough to jolt me off that path and I ricocheted from major to major, trying Latin, classics, and government before settling on history because I had the most credits in that subject and found it interesting. In the spring of 2003, I had a devastatingly bad semester where I went through three solid months in a severe depression, skipping almost all my classes. I would have flunked out of college entirely, but a sympathetic academic dean helped me secure a retroactive medical withdrawal. It was a drastic departure for someone who had always been an overachiever in school and extracurriculars, and my equally overachieving FBI agent parents were appalled that their star kid had nearly failed. To attend Smith again, the college required me to take a full semester of classes at another university to prove I could handle the academic load.

I settled on UMass Amherst, which had extension courses available for nondegree seekers. My parents told me they would not help me financially since I had wasted a semester of their tuition money. Failing is not something my family does, so their anger didn't surprise me. I still wanted to make my parents proud, a compulsion baked into my being from a young age. In response to being cut off, I found two jobs to cover my tuition and living costs. One was in the evenings and on weekends, working at a local retail store that sold every type of product conceivable for dogs. The other was weekday mornings at the local paper, the *Daily Hampshire Gazette*. That work schedule meant I could only take classes that began after 1:30 p.m., or that met only once a week for four-hour blocks. Since this seemed like my only shot at academic redemption, I decided to take classes that would fully capture my interest: Studio Drawing, Poetry, Indian Peoples of North America, Literature of Travel and Place, and Introductory Geology.

The geology course was revelatory. It met on Wednesday evenings, from 6 to 10 p.m. The professor spent the entire four hours drawing figures and scribbling nearly indecipherable notes on an overhead projector with his back to us. His monotone delivery put most of the students to sleep at least once during each session, but I was riveted. Here we were, in a depressingly institutional windowless classroom from the 1960s, learning how mountain ranges were formed, how volcanoes spew rivers of lava, and the vast history of the Earth as seen through the scope of geologic time. Billions of years of our planet's past were laid out for me to explore. I understood my favorite lines from William Blake's poem "Auguries of Innocence" in a new light:

> To see a World in a Grain of Sand
> And a Heaven in a Wild Flower
> Hold Infinity in the palm of your hand
> And Eternity in an hour

Every grain of sand has its origins deep within the Earth's molten interior, and each rock tells the story of its formation and ongoing destruction. A whole language is hidden within the science of geology and this taste of deciphering it opened undreamt-of depths of understanding about our world to me. How had I overlooked this entire field of study? How had no one told me that traveling the world and discovering its secrets was still a career option in the twenty-first century?

I finished the semester on the dean's list and was readmitted to Smith. With only two semesters remaining before my graduation I didn't have time to change my major again, but I enrolled in as many geology courses as I could. I left college believing I would have to be content with my too-brief experience of earth science from those few undergraduate courses.

Needing sunshine and warmth, I fled Western Massachusetts for Phoenix, Arizona, immediately after graduating. I worked as a veterinary technician, in the Arizona State Archives, and as an insurance claims investigator over the course of a few years before arriving at a crossroads. My job options and ability to contribute to society seemed limited without an advanced degree. I took the LSAT examination for potential application to law school, but while my score was high enough to earn admission to a good program my heart just wasn't in it. I thought about how much I hated working indoors, sitting at a desk day after day, and remembered how deeply I had loved the single geology research expedition I had participated in as a naive undergraduate. That prompted me to research second bachelor's degree options, in the hope that a geology degree would let me work outside—at least some of the time. Going back to school as an undergraduate seemed like torture, but it would be better than spending three years locked within the bowels of a law library only to emerge as an attorney.

Around that time, my serious boyfriend proclaimed he was moving to Los Angeles, ostensibly to become the world's greatest film director (spoiler: he did not). I looked for geology programs around L.A. and I stumbled across a master's make-up program at California State University, Los Angeles. Students with a degree in any field could enter this track. While enrolled as a graduate student and taking graduate-level classes, I would make up missing undergraduate-level foundational courses. The program might take longer than a traditional master's program, but it seemed like it might work for me. I had already taken calculus and chemistry, so I would only need physics and a few introductory geology courses to keep pace with regular graduate students. In March 2010, I graduated from CSULA with my Master of Science degree, complete with the signature of Governor Arnold Schwarzenegger confirming my official status as a geologist.

From my thesis research on an undersea Hawaiian volcano, my work has taken me to the peaks of the Peruvian Andes, the rugged mountains of cartel-controlled Sinaloa, Mexico, the austere grandeur of the Australian Outback, the endless expanse of the Mongolian steppe, the wilds of the Grand Tetons, and throughout California's own brutal Mojave Desert.

On most jobs, I've been the only woman present. I've hauled my own gear, handled injuries and emergencies with professional calm, and navigated the oceans of testosterone by learning to cuss as effectively as an Australian coal miner. I saw how the narrowing scope of scientific research can leave scientists missing the forest amid the attention they pay to the trees. Over and over, I watched as lesser-qualified men were promoted above the few women who were working their asses off to deliver solid science. I believed I was personally immune to the effects of misogyny, since I was raised playing sports on boys' teams and saw myself as someone outside sexism's stifling grasp. I was wrong, but it was a few years before I learned to recognize the insidious ways women, people of color, and those with disabilities find doors to scientific opportunity closed in their faces.

I've come to expect people who do not understand the rigors of real scientific fieldwork comparing my field experience to the errant wanderings of people like the protagonist of Jon Krakauer's *Into the Wild*, cheapening the value field science adds to our world.

When I taught college courses, I witnessed a student overwhelmed by the prospect of an oceanography field trip on a boat off the coast of Long Beach. He was twenty-two, had spent his entire life in Los Angeles just 12 miles from the ocean, and had never visited the beach because he spent every day working at his parents' Mexican restaurant. College was his ticket out of that life, but this required science course that was not even part of his major upended his world. His story was far from unique, and I slowly realized that while historically underrepresented people are earning more and more of all degrees awarded in science, they remain largely locked out of jobs or receiving grants in the sciences. Just going on a field science course to learn research skills can cost thousands of dollars, which is an enormous barrier to many students. How can you learn to see the forest for the trees if you cannot afford the price of admission to the wilderness?

Wanting to knock down those barriers to science is what drove me to found Blueprint Earth, which focuses on producing valuable environmental research while providing real-world educational opportunities to students who would not typically have access to scientific field research expeditions. I vowed that the fieldwork experience we gave students would be offered at no cost to them. My board of directors helped crowdfund $14,000, and that was our launching platform. If science could not be made available and accessible to everyone, we humans were doing ourselves a disservice as a species. Dealing with the challenges of a rapidly changing climate and exploding population growth demands an all-hands-on-deck approach—not just the talents of those who can afford education at elite schools or who look like our clichéd representation of a scientist. My hope was that I could use the Explorers Club to connect students to opportunities and amplify Blueprint Earth's environmental science work.

▲

On this March weekend, explorers from every corner of the planet were gathering in New York for several days of festivities surrounding the Explorers Club Annual Dinner. The evening before the dinner was reserved for the new members reception and a general cocktail reception. Members can bring guests to mingle with some of the brightest stars in the exploration community. If a guest is accompanied by a club member, the whole club is open for them to visit. Club staff give guided tours of the building highlights that are usually off-limits, including a stuffed sperm whale penis, a collection of maps commissioned by Napoleon

Bonaparte, and a club flag that has journeyed to both the bottom
of the Mariana Trench and the summit of Mount Everest. It is
an experience rich in history and tradition, the kind of evening
dreams are nourished on.

Wandering the corridors of the Explorers Club headquarters,
the musk of old leather, taxidermied animals, and the colognes of
explorers past and present filled my nose. Men with white fringes
of hair surrounding their bald crowns brushed past us, conversing about Nepal and cheetahs. The occasional woman strode by
with the practiced ease of someone accustomed to wearing heels,
drink in hand, and bedecked in spectacular necklaces made of
giant megalodon shark teeth or intricate colorful beadwork,
surely crafted by an indigenous tribe member from somewhere
in the Amazon basin. They were clearly comfortable with an exaggerated version of cocktail attire, and I was quickly learning that
everything in the club was outsized—just like the tales of explorations I overheard as we walked. Overhead, the wood-paneled
hallway ceilings ceded to the intricate white moldings of the
much-higher room ceilings. Rich carpeting kept conversational
noise to a low hum and meant that Carlos and I had to strain to
hear each other if we strayed more than a few feet apart.

Waiters in black vests over crisp white shirts hurried past the
ferocious polar bear, pushing small carts packed with bottles of
chilled wine toward bars set up on either end of the second floor.
People smiled politely at us, but it was obvious that most of the
guests had longstanding friendships and were about double my

thirty-four years. Since many geologists I worked with were of the same vintage, I was prepared to be a bit of anomaly. Still, I was hopeful we'd find some people closer to our ages. We made our way to the bar and acquired drinks before tagging along with a tour group to learn more about the club's Pandora's box of lore.

The tour was engaging and finished on the top floor of the building in a cavernous room with a peaked ceiling and exposed wooden support beams. The walls, tables, and even the sloped ceilings were covered with artifacts. The pelt of a man-eating tiger killed by a club member in the interest of public safety was one of the objects on the ceiling, its glassy eyes overseeing the meanderings of the visitors below. This was the club's "trophy room," a holdover from the days before conservation and science replaced trophy hunting and adventuring as the club's focus. I moved to the far end of the room, inspecting the tusk of a mammoth hanging from the ceiling beams. A now-familiar deep voice called my name questioningly. I looked around to find Bill Steele in the flesh, flanked by Jim Smith.

Bill wore a vest over a button-down shirt, his full head of white hair crowning a buoyant face and blue eyes charged with enthusiasm. He had a neatly trimmed beard and mustache, also white. His handshake was appropriately firm for a man who spent a good portion of his life deep underground, scrambling through a cave featuring a section called "The Torture Chamber." Jim looked to be about the same age as Bill, somewhat heavier-set but with the same strong grip. He had a similarly impressive beard shot through with streaks of gray. It was easy to envision them rappelling down rock faces in caves that would terrify most people. They both seemed genuinely happy to see me, offering congratulations on my acceptance to the club. I thanked them, noting that I wouldn't be there without their help.

We talked for a while, and Bill told us what to expect at tomorrow's dinner. He would be wearing native dress, he said, and I wondered what that meant since I thought the dinner was black tie. According to Bill, many members come dressed in traditional attire from the places where they perform their fieldwork or from their own personal heritage. Since Bill looked to be of European extraction, I asked what he would be wearing. He explained he would be wearing the clothing of a *curandero*, the Oaxacan healers who help assure his cave expeditions are safe and successful by communing with the spirits of the cave. He wanted to be in the right frame of mind to approach the curanderos when he met them in just a few weeks.

Listening to Bill and Jim talk about expeditions past and future, I noted how stereotypical their names were for explorers: Bill Steele and Jim Smith. They could have stepped from the pages of an adventure novel, and yet their camaraderie was genuine and warm. It seemed that a generational shift was in progress, since I was far outside the traditional definition of a typical explorer. When they learned that Carlos was Blueprint Earth's cofounder, and present on our research expeditions as a data management expert, they urged him to apply for membership. He was one of just a few non-white people I saw in the building, yet they were eager for him to join the club's ranks. This affirmed my impression that the club was actively moving past its troubling colonial roots, and that Bill and Jim were good guys.

After we parted company with the cavers, we found a quiet corner of the building to talk. I was enraptured with the club's history and potential—despite the copious amounts of taxidermy still present at headquarters—and wanted to brainstorm about how we could encourage Blueprint Earth's student researchers to apply for membership. Most of our students are women, or people of color, or from low-income backgrounds, and many have a combination of those traits. At the heart of my enthusiasm was something foreign yet intoxicating. Unable to quickly identify the feeling, my mind raced with the possibilities awaiting the next generation of explorers.

For the first time, I had found a group of people who understood my burning desire to dig into the beating heart of our planet. I had never been content just to read about distant lands and other peoples' work uncovering the secrets of the universe. Joining the effort to understand our world spoke to the child that still lived inside me, the kid who grew up on the novels of Jules Verne, Robert Louis Stevenson, and Jack London, and devoured the televised science lessons of Mr. Wizard. The shadow of the *Challenger* and *Columbia* disasters underscored how fragile our exploration efforts could still be. Armstrong's small step had shown humankind that giant leaps are possible if we enlist our collective strength and intellect in the service of something greater than ourselves. Here was a group that not only dreamed of the possible but set out to prove what possible really was. A group that recognized the role of science in illuminating the darkness of the human condition, of shedding light on the theoretical by turning it into the concrete. A group where I belonged.

I had earned acceptance into this extraordinary society not through being the first to conquer some lofty peak or make contact with an isolated tribe. I was here because today's exploration belongs to scientists. It is the domain of the curious, for whom no amount of knowledge will ever be enough. Scientists are people whose curiosity drives our life's work. My curiosity shoved me headlong into a quest for truth, for objective reality—often in the face of danger and at great personal risk—and has never once allowed me to become complacent or overly confident.

That constant, unbridled yearning for understanding our place in the universe is what binds explorers and scientists. T. S. Eliot's poem "East Coker" ends with almost a rallying cry for explorers, both modern and historic.

> Old men ought to be explorers
> Here or there does not matter
> We must be still and still moving
> Into another intensity
> For a further union, a deeper communion

Through the dark cold and the empty desolation,

The wave cry, the wind cry, the vast waters

Of the petrel and the porpoise. In my end is my beginning.

Exploration is not about the individual explorer, just as science is not about the scientist. Exploration and science are our birthrights as humans, and as I have found my life's purpose through exploring our world, my experiences belong in the great catalog of exploration. Every life is an exploration, a testing of hypothesis after hypothesis as we seek to eliminate uncertainty, adjusting and learning, growing and changing, and carving our own trail, however unique or unlikely it may be. We are the explorers, and it is through our exploration that we finally come to know ourselves.

# Into the Valley
# of Death

MARCH 2004, CALIFORNIA

THE PLANE JOUNCED through a pocket of turbulence, shuddering. The overhead bins groaned. I wiped my palms across my jeans, ready for the five hours of imprisonment to end. With my forehead pressing into the cold plexiglass, I watched as snow-capped peaks morphed into harsh outcrops of sandstone and dark lava rock divided only by the occasional meandering of a tired southwestern river. I was midway through my second geology class—complete with coveted National Science Foundation funding—and we were now making our initial descent into McCarran International Airport, Las Vegas. Death Valley is not the spring break destination of choice for most college students, but in just a few hours eleven aspiring scientists would be introduced to the devastating beauty of the hottest place on Earth.

I was intensely nervous. The name of our study location evoked a visceral sort of fear, which has biblical roots for those raised in any manner of Christianity. Furthermore, the other students, most of whom were majoring in some type of natural science, were experienced campers, whereas I had never camped before. The prospect of nearly two weeks of long days in the desert sun with no running water—no toilets—was more than mildly concerning to me. I had resolved to grin and bear it as long as humanly possible while simultaneously praying that we would need to make a run to civilization for gas, or food, or perhaps even some sort of minor medical emergency. This plan had several flaws, but I was an optimistic history major entering a geologist's world, and the lure of a free (and educational) trip to California was too strong to be offset by concerns about bodily functions.

I was a fish out of water in more ways than one. The lone student from Smith College in a group of Hampshire College students, I had taken advantage of the Five College Consortium benefits offered by our Western Massachusetts schools and registered in a Hampshire class. Among the Five Colleges, Hampshire is known as the hippie school. (It is rumored that a student once graduated after completing an undergraduate thesis project on the frisbee.) Smith College, on the other hand, had a bit of a holdover reputation from its years as Yale's sister school, with some wrongly dismissing it as a finishing school for young ladies. With over 25 percent of the college openly self-identified as lesbian—not counting the substantial bisexual, pansexual, and other differently oriented students—the Smith of 2004 was a far cry from the pearl-clutching days of old. However, compared to my Birkenstock-clad, patchouli-drenched Hampshire classmates, I felt like the love child of Footlocker and Hot Topic: my sporty look more suited to making a visual statement than surviving in the wilderness.

Prior to the March expedition, our class met a handful of times on Hampshire's campus. As a lifelong pursuer of straight As, I was still uncomfortable with the Hampshire system of not assigning grades for completed coursework. Our professor, Steve, assured me that his long-form written evaluation of my work would be enough for Smith's academic overseers, but I had my doubts. Fortunately, the official course title was enough to negate my grade preoccupation: Evolution of the Landscape. Why do the Appalachian Mountains exist? How can we tell that this valley was an ancient lakebed? Where is the evidence that glaciers once covered Manhattan? I hoped that the answers to a lifetime of burning questions would be revealed in this course. The promise of a fully funded trip to California to assist with dating prehistoric lakeshores sealed the deal. I would have the opportunity to help add to the body of knowledge about our planet. Game on.

The role of the geologist is like that of a detective tasked with re-creating a scene that played out before their arrival. In technical geology terms, our class was one in geomorphology. The focus

of geomorphology is to examine how landscapes are formed, and
that was the purpose behind our professor hauling a group of
novice scientists to the decidedly nontourist wilderness of Panamint Valley. Unlike in other fields, when geologists say that they
need to uncover evidence to prove a hypothesis, it is likely they
mean that literally.

We came to Panamint Valley to dig.

As the plane bumped to a landing on the shimmering
Vegas tarmac, I reminded myself that most planes do not burst
into flames upon touchdown. The relief of stretching my legs
on solid ground was dampened by the omnipresent odor of
tobacco. The slot machines flashed enticingly, but our group
was on a mission. Steve and Jason, the Hampshire microbiologist we would be helping in addition to our geology duties, set
a determined pace as they parted the sea of alternately hopeful
and heartbroken airport gamblers. We trailed along in their
wake, certainly the only group of college spring breakers arriving in Sin City on a quest for chlorine-36 isotopes rather than
ethanol-based drinks.

▲

The plan was to camp that night somewhere between Las Vegas
and Death Valley. We piled into the standard-issue white rental
vans, making small talk and marveling at the stark exposures
of warm-hued rocks jutting skyward to the west of the airport.

The contrast to the riotously vegetated part of Massachusetts we had just left was jarring. My mind, intoxicated with the delicious revelations of classroom geology, was reeling from the effort of identifying rock types while also handling conversation and jet lag. Dusk fell behind us as we chased the setting sun on Highway 160, winding through the Spring Mountains and onto the sweeping expanse of the Pahrump Valley. Redolent desert air streamed through the open van windows, heavy with dust and memories that belonged to another time. Just as the monotony of the drive began to win out over excitement, Steve spun the wheel right and we sat up as the van bounced off the highway and onto a blink-and-you'll-miss-it washboard road.

Steve, abstractly factual as always, informed us why we were suddenly tooling down a road more suited to ATVs than Ford Econoline vans.

"We'll be camping here tonight!"

"Where's here?" A student queried, speaking for the group.

"I don't think it has a name," Steve responded.

Not wanting to reveal the true depth of my camping ignorance, I stayed silent. But I had questions—several. Could someone help me set up a tent? How do you set up a tent in the dark, anyway? Did this unnamed campsite have bathrooms? Would there be other campers?

As the van clattered along for what seemed like several miles, the remoteness of our situation became apparent. I could not make out a light, save those coming from Jason's van behind us. The road deteriorated until we made another hard right onto a track that could only generously be described as a road. Mountains loomed on three sides of this nameless corner of Nevada. Steve parked the van abruptly and dust swirled, smothering the headlights.

"We're here!" he chirped. "Since it's late, let's not set up tents. The weather's perfect for open-air camping."

The other students gave murmurs of approval, while I entered a minor panic. I had clung steadfastly to the notion of the tent as an essential barrier against the unfamiliar nighttime dangers of the Nevada desert. Now I would need to camp, for the first time in my life, without even that thinnest layer of protection. I swallowed hard and resolved not to burden the group of Hampshire-ite earth children with my camping virginity.

Long after the rest of the group was snoring, snuffling, and otherwise drifting through dreamland, I stared upward into the endless bowl of desert sky. The light pollution from Las Vegas provided enough ambient glow to discern some of the darker rock layers in the mountains closest to us. Several sharp pebbles were constant reminders that I was indeed sleeping on the ground with nothing between me and the rocks I was fascinated with but some cheap polyfill and a few layers of nylon. I pondered the existence of scorpions and recalled with dismay that my father had once brought one home from a work trip to Phoenix—inside a glass paperweight. I hoped scorpions were not karmically connected. The stars spun, sliding across the sky around the axis of the North Star.

When dawn began to peer over the mountains, I realized that I had not been lying only on rocks. At least thirty scraps of round metal joined to brightly colored plastic cylinders littered the ground next to and under my sleeping bag. I picked one up, squinting at the number 12 stamped into the metal.

"There's not much to do out here other than shoot at stuff," said Steve, interrupting my study of the spent shotgun shell. I dropped it like it was full of vengeful scorpions and attempted to look knowledgeable about shell casings.

My classmates were sitting up, rubbing sleep from their eyes and fumbling for toothbrushes. In a hurry to distance myself from the spent shells, I squirmed out of my sleeping bag, stretching to greet the day. The morning air was light and crisp, its chill fading with each passing moment. A stop at the local store for eleven days of groceries awaited us across the valley below.

▲

After an hour and a half of driving, we reached the outskirts of our destination. My initial view of Death Valley is seared into my memory, nature's first permanent brand upon my adult psyche. Since the internet was still in its relative infancy, with Google five months away from its IPO and subsequent creation of a new verb, I had not done any research on what Death Valley National Park looked like. Even if I had, it would have been underwhelming compared to seeing it in person. I do not regret that my first impression was made there at Zabriskie Point, surrounded by a group of intelligent, nature-loving, experienced outdoorspeople who were every bit as awed by the experience as I was.

The Earth's skeleton was heaved up before me, with its scars and secret mysteries ripe for exploration. Modern washes knifed through ancient lake sediments, yellow and rust-tinted and weathered by eons of water and wind. Dark lavas rest on top of the remains of those prehistoric lakes, reminders of the region's more recent—and more violent—past.

I sat against a rough stone wall, its solid mass grounding me against the exponentially expanding flood of prehistory coursing through my mind. Steve was explaining to the group that the shades of tan, brown, red, green, and white were due to the interaction of hot geothermal waters with minerals in the rocks. I barely registered his words, struggling to take in the sliver of Death Valley visible from our vantage point in the Amargosa Range of the Black Mountains.

Already, I was captivated by the names of this place: Death Valley, Amargosa, Black Mountains, Funeral Mountains, Telescope Peak, Panamint, Badwater, Zabriskie, Teakettle Junction, Racetrack Playa, Marble Canyon, Furnace Creek, Darwin Falls, Artist's Palette. I mouthed the words to myself, savoring their unfamiliarity while trying to match them to the pictures they conjured in my imagination. Visions of well-adapted natives and callow pioneers threatened to waylay my thoughts entirely, and it was with effort that I reined my mind back to geology.

Death Valley itself barely peeked out between the tall sable hills to the south and the lower buttery hills to the north. The size of the valley is not impressive from Zabriskie Point. What is impressive, though, is the obvious relief between the valley floor and the mountains of the Panamint Range to the west. My brain was churning through the materials we had covered in class—the process of the Earth's crust thinning and pulling apart, creating valleys between mountain ranges through the western United States. The Basin and Range geologic province is studied around the world. It felt as though we were standing inside an illustration from a Geology 101 textbook.

As we set off on a short hike through the area below Zabriskie Point, Steve cautioned us against the dangers of not carrying enough water and of splitting off from the group. He told us about a hiker who had gone out with only a liter or so of water, rapidly became disoriented and dehydrated, and died—right in the area where we were hiking so casually. As a novice hiker, I made a mental note to stick close to Steve, Jason, and the most experienced students in our group. I trailed behind Steve's lanky form, taking two strides for each of his. We had yet to set up a proper camp, but I had already survived a night in the open air with scorpions and shotgun shells. I was not about to die before getting a crack at doing actual field science. I was (inordinately) proud and determined.

▲

The initial days of the trip eased me into the whole outdoor science experience. Our first official campsite was at Furnace Creek Ranch in Death Valley, which is where most national park visitors stay. With spots available for both tents and RVs, generator sounds crowded the night air, but the campsite had nice sinks, flush toilets, and even showers you could pay to use. Since several of the Hampshire students had resolved not to shower for the duration of the trip, I decided that this adjusted view of personal hygiene would suit me, too. Patchouli oil was still a no-go, but friendships were forming during our late-night discussions of constellations, Native American mythology, and general campfire stories. Growing up in Colorado, I had received a good amount of education in the cultures of various Plains peoples, as well as a smattering of knowledge about Desert Southwest, Mexican, and Central and South American peoples. Sharing that information gave me a sense of having contributed to the group. I was still concerned about potentially needing help throughout the trip, so it eased my worries to feel like I could give something back.

The science so far had been limited to helping Jason take microbiological samples from Badwater, one of Death Valley's most famous—and famously protected—sites. When our Death Valley trip took place, Badwater was known as the lowest point in the Western Hemisphere, at 279 feet *below* sea level. The recognition that Laguna del Carbón in Argentina is the lowest point in both the Western and Southern Hemispheres at −344 feet was still a few years away, so it was with a special feeling of reverence that I set about my task of sampling in the name of biology. A plaque affixed high on the mountain to the east of us provided a visual marker. It bore a thick whitish line and the words "Sea Level." Perspective.

Jason explained that we were looking for halophiles, which are a type of organism called an extremophile that exist in water that is too salty for most other life. Extremophiles are capable of thriving in environments known to be hostile to human life. This means they flourish in extreme cold, heat, salinity, alkalinity, acidity, or a host of other conditions that would kill most creatures. Many are tiny microbes, but some are larger, such as the colossal squid, which is a polyextremophile since it's both a barophile (survives in high-pressure environments) and a cryophile (survives in cold environments). Scientists are interested in extremophiles since they provide clues as to how life began and evolved on Earth, and as to what extraterrestrial life-forms may look like. For example, the teddy bear–like, eight-legged, wormish tardigrade is an extremophile's extremophile. It can withstand radiation, starvation, dehydration, extreme pressures and temperatures, and oxygen deprivation. Finding something like a tardigrade in space is much more likely than finding a little green humanoid, so understanding the variety of extremophiles is a priority for scientists.

Badwater is named for the brackish water that fills the eponymous Badwater Basin, located near the southern end of Death Valley. It has a high concentration of salt, which makes it unappealing to drink, but exactly what Jason was seeking. The water found there bubbles up from an underground spring and doesn't flow out to any rivers or the ocean, forming what scientists call an endorheic basin. Jason explained that the extremophiles we might find at Badwater would be microscopic. He was energized, his dark eyes dancing as he described how to use our tools to push algae and the thin crust at the bottom of the pond into our sample tubes. We were to lie down on the wooden observation deck to reach into the 8 or so inches of brackish, federally protected waters. In the interest of fairness, we alternated sampling and note-taking with partners. There was enough scientific glory to go around, fortunately.

As I delicately positioned my chest on the splinter-laden boards, I resolved to sample so carefully that the tiny aquatic creatures populating Badwater would not even notice a few kidnapped extremophiles. With my face no more than 6 inches from the water's surface, the smell of salty, stagnant water invaded my sinuses, causing my eyes to water. I blinked and pushed both gloved hands below the surface, into this environmental mystery that also happened to be a major tourist attraction.

Scrape scrape scrape. Tiny whirlwinds of black-green debris swirled up from the bottom of the shallow lake, obscuring the russet sediment at the bottom. I was fervently hoping that at least some of the algae had found its way into the sample tube when I heard a shrill voice that seemed to rise ten decibels with each word.

"What. Are. You. Doing?!"

I glanced back at my interrogator, a task made nearly impossible by my position face down and up to my biceps in the mysterious waters. Her round, reddish face was etched with disbelief and the beginnings of anger. Her voice continuing to rise, the woman repeated the question to my now-terrified sampling partner. I watched something black and shiny and vaguely slug-like undulate away from my scraping, and wished that I could undulate away, too. I could either stop sampling, rendering this specimen useless, or keep working and act like English was not in my wheelhouse. Then I felt the wooden boards vibrate beneath me. A second later, salvation arrived.

"It's ok, ma'am. We have scientific permits. We're conducting a study of halophiles here at Badwater. We're scientists," Jason said, the technical terms adding credibility to his explanation.

I heard my partner exhale, her sigh matching my own. I had taken a grand total of one science course before this semester, and Introductory Geology did not exactly confer scientist qualifications. Still, if Jason was going to include us in his efforts at mollifying agitated witnesses to our scientific desecration of hallowed natural ground, I was for it. Scientific permits abruptly took on special significance in my life. Who knew what other feats of intellectual inquiry could be open to me if I had a scientific permit? The possibilities seemed endless. I managed to complete my sample and sat up, dizzy. I handed the sample tube to Jason, who had successfully pacified the now-embarrassed tourist and sent her wandering off into the endless sea of salt crystal hexagons that reach westward into the center of the valley. Photo opportunities were much more interesting than watching scientists dig around in muck and slime, anyway. Keeping good relations with the public seemed to be an important lesson. Working in a national park meant inquiring eyes following our every move and knowing how to quickly communicate both the legality and significance of our work looked to be an asset.

"Does it usually work that way? Just say you're a scientist and they let you do whatever you need to do?" I asked Jason.

"In the United States, yes," he responded, grinning. "Other countries, not always."

The thought of doing science in other countries had not yet crossed my mind. I dried my hands and took a moment to snap a picture of the perfect mirror image of a mountain captured on Badwater's glassy surface, pondering what it would be like to work in places that I had only read about. Just two days before, Badwater had been that sort of place—a place that only existed in encyclopedia articles and stunning *National Geographic* spreads. Now I was covered in extremophiles, or at least mud, and stinking of saltwater in the hottest place on Earth, eagerly anticipating getting to the best part of the trip—the geology. I was already trying out my new favorite phrase, turning it over and over in my mind like a new toy. *It's ok, I'm a geologist.* Science has the power to unlock the secrets of the universe, and I was ready to get cracking.

▲

The sights of Death Valley are myriad and stunning, but the experience close to the paved roads is still tourist-friendly. Helpful signs provide details about park geology, biology, and history, and major points of interest have basic restroom facilities. Despite its forbidding name, Death Valley is, in fact, vibrant and full of both animal and human life. From enormous shifting sand dunes to marble slot canyons scoured smooth by floodwaters, the beauty and brutality of the park is obvious to visitors. Our group took in a few noteworthy sites before launching out beyond the national park boundaries and into the wilds to the west of Death Valley.

Panamint Valley sits about a thousand feet higher in elevation than Death Valley. Waters from prehistoric Panamint Lake spilled over into its more famous neighbor during the much wetter geologic period before the present, the Pleistocene, when glaciers covered vast portions of North America. Both Death and Panamint Valleys were part of a larger system of lakes and rivers during this period. The evidence for this interconnected series of lakes is recorded, as most significant climatic events are, in the rock record. It was up to us to get the rocks.

Our faithful white van, now beginning to smell decidedly of unwashed college student, chugged up the roughly 5,000-foot-high Towne Pass. Signs warning drivers to turn off their air-conditioning dotted the shoulder of the road. This stretch was so steep, and frequently so punishingly hot, that even government signage hinted at its reputation as a car-killer. We passed what looked like a film crew standing on the side of the road behind three cars wrapped in brown covering, with nothing aside from their windshields and windows left clear. Their makes and models were completely obscured. Steve explained that Death Valley is popular with automakers, who use the area's extremely high temperatures to test vehicle performance.

Steve had the look of a guy you could trust with your life in a wilderness situation. Like most geologists, he wore the standard-issue grizzled outdoorsman beard. I am never sure if this is a conscious geology fashion choice, or simply an acknowledgment that shaving while doing fieldwork in remote areas is just not going to happen. Steve's mountain man visage was softened by the constant sparkle in his eyes, which burned even brighter whenever someone asked a particularly good question about rocks. Also, par for the geologist course, Steve's fashion sense looked like a mixture of L.L. Bean catalog and thrift store. His neutral-toned pants concealed most dirt, and his hiking boots were scuffed. But despite his rough and ready look, Steve's music preferences were cool enough to impress a group of early twenty-somethings. We cruised the sinuous, pockmarked blacktop down into Panamint Valley to the tunes of Cake, Steve gaining points for singing along to "The Distance." We joined in.

*"He's going the distance / He's going for speed"*

We stopped for a roadside lunch of tinfoil-compressed peanut butter and jelly sandwiches, which tasted inexplicably better

when consumed while listening to Steve tell the story of Panamint Valley. I conjured a mental image of a beautiful temperate
lake, like those of modern New England, and tried to layer it over
the stark, rocky emptiness spread out below us. With some effort,
I could almost will it into being. To me, the effect was better than
a movie, since I was sitting in the environment while imagining
its transformation.

I sat next to Anne Marie, a Hampshire student who was
curious, outdoorsy, and who I'd never seen wearing anything
other than Birkenstocks prior to this trip. She and I discussed
the various discomforts we were noticing after a few days away
from civilization. Some of the women on the trip had mentioned
to us, in confidence, that they were having their periods. I was
horrified. Our male trip leaders had failed to mention this potential issue during our pre-departure discussions, and as a rookie
camper, I had not even considered the logistics of backcountry
feminine hygiene. Minor blisters were my biggest concern at this
early stage of the expedition, and for that I thanked whichever
merciful geology god(dess) had spared me a major complication
on my first camping trip.

I was, however, growing a bit worried about the rapidly
approaching test I would have to pass. Up until this point, we
had access to park service toilets at least twice each day. After
we reached our Panamint Valley campsite, we would not see
another porcelain fixture for well over a week. The prospect of
temperatures over 105°F seemed like an interesting challenge to

surmount. Snakes, scorpions, and heatstroke appeared somehow reasonable and fine. Figuring out how to defecate successfully with nothing to sit on? That was a puzzle. I resolved to ask Anne Marie about the mechanics if the situation became critical.

After lunch we continued down to the center of Panamint Valley, where tens of thousands of tiny yellow flowers broke the tan desolation of the valley floor. Steve swung south onto a road that had spent years waiting for repairs, its patchwork of cracks regular and gaping. He changed CDs, popping in an unfamiliar but decidedly appropriate Tom Petty album.

*"You belong among the wildflowers / You belong in a boat out at sea"*

Petty's reedy voice matched the pace of the van skimming across the barren desert, dotted with unexpected bursts of yellow flowers. The music coursed through me, washing away my lingering doubts about belonging with this group who seemed so comfortable living and working outdoors. Sure, I was oblivious to many things that my classmates took for granted. Headlamps? Never heard of them, but they did look ideal for hands-free illumination at night. Sock liners? I had not realized that those even existed, let alone packed any. REI? Is that a band? My inner awkwardness and shame began to fall away, and I felt myself sinking into the essence of the desert.

We passed alluvial fans, the fan-shaped deposits of boulders and rocks that funnel down out of mountain canyons during rain and rock falls, spreading across the flatter areas below. They can be tens of miles long, and Panamint Valley has some impressive ones. The second song of the Tom Petty album began, moving quickly into the chorus.

*"But let me get to the point, let's roll another joint"*

It fit the mood perfectly, but I will admit to being shocked. My parents were FBI agents, and my sheltered mind firmly held that people with PhDs never, ever did drugs, and certainly would never condone drug use in any way. Steve, contrary to expectation, sang along. I glanced at my Hampshire traveling companions, looking for any sign of surprise. With none evident, I recalled that Hampshire's reputation as the hippie school of the Five College Consortium came with the implicit understanding that most of its students were relaxed about drug consumption. I decided to relax, too. Hell, we were literally in the middle of the most obvious expanse of nowhere I had ever been. Tom Petty and his melodious musings were going to see us through.

▲

Our temporary home was 2,000 feet above the floor of Panamint Valley, three-quarters of the way up one of the gradual slopes of an alluvial fan on the valley's western side. A fifth-wheel trailer and two pickup trucks were waiting for us. Steve's colleagues from schools out west had beaten us to the site, and we shook hands with an even more grizzled geologist who resembled a tommyknocker, or a solidly built leprechaun in his seventies, with a white-yellow beard that engulfed the lower half of his face and seemed to cover most of his chest. The top of his head came just to my chin, and he had on worn dark jeans barely held up by haggard red suspenders. His flannel shirt had more than a few roughly patched holes, and an aura of what I assumed was chewing tobacco permeated the air around him. His name was Roger. I was disappointed it was not Rumpelstiltskin.

When Steve mentioned we were meeting his collaborators, my classroom-oriented assumptions had kicked in, and I was expecting someone closer to my sweater vest–clad history professors. The other scientist Steve introduced to us fit my impression of what a scientist would look like. Bespectacled and sensibly dressed in gear from leading outdoor companies, Fred the hydrologist seemed more interested in talking to Steve about tomorrow's research plan than getting to know our pungent group.

We set to pitching tents and preparing our campsite. Part of this work involved deciding where to designate as the bathroom area. This was a bit of a challenge, as fresh alluvial fans like the one we were camped on were covered in boulders but decidedly lacking in trees or shelter of any kind. One of the guys solved the dilemma by walking away from camp until a slight depression hid him from view. We called him back and agreed that the general direction of "that way," would now be our restroom facility. Steve sauntered over and asked where we had chosen. We pointed and he strode off carrying a small shovel, a roll of toilet

paper peeping out of one pocket. A single square of paper flut-
tered jauntily as he walked.

While we had received no instruction in the mechanics of
going to the bathroom in the desert, Steve had clearly told us that
used toilet paper cannot be buried—there is so little rainfall that
it could be decades before even small amounts of paper break
down. Instead we would need to burn it at the scene or pack it
out for later disposal. Everyone carried lighters and plastic bag-
gies as standard equipment.

Dinner was relaxed but heavy with anticipation for the next
day and the beginning of the real heart of our scientific mission.
Steve and Fred did most of the talking, relating stories of their
work in Antarctica. My mind wandered, thinking of what it
would be like to be paid to travel the world, learning and sharing
knowledge. Below us, Panamint Valley was shrouded in dark,
with not a single light visible from our camp. It was as if we were
the only group who knew this place even existed. I felt a deep
but comfortable solitude, even surrounded by more than a dozen
people sitting around a cozy fire.

▲

Before dawn, I crawled out of my tent in a quiet frenzy, desperate
to get out of the line of sight to pee. I booked it across the rubble
of the alluvial fan, toilet paper and lighter clutched tightly. After
business was finished, I was watching tongues of fire devour my

used toilet paper in the purple-gray pallor of the morning when a low rumble from the south shook the air. I dropped the toilet paper, now mostly just carbon, and rubbed out the flames with my shoe. The noise was growing, and I could feel my chest begin to vibrate ever so slightly.

Even with my geology knowledge in its infancy, I had investigated California's famous earthquakes, and had read that they are often accompanied by a low rumble. I tensed, waiting to feel the tremor in the ground that means the earth is shifting. Instead, a small silver shape tearing across the floor of the valley caught my eye. It was a fighter jet, and it was flying *below* us, winging along just a few hundred feet above the floor of the valley, distinctly out of place and yet somehow perfectly at home. Marveling at the military's ballsy use of the empty space of Panamint, I made my way back to camp to put my daypack together.

▲

Our scientific mission was to search for ancient lakeshores that could be dated using a technique known as cosmogenic radionuclide dating. This work would help establish a more concrete record of when the Earth's climate was different than it is today, which adds to the understanding of how life has shifted and evolved over hundreds of thousands and even millions of years. What is now a brutal desert was at one time covered in a shallow ocean, and areas now inhospitable to most creatures used to be home to lumbering dinosaurs, giant ground sloths, or perhaps creatures whose remains were never preserved by any rocks and are now lost to time.

Cosmic rays constantly bombard the Earth and interact with atoms in our atmosphere as they rain down. These interactions can cause parts of the atoms to become dislodged, forming different elements, or new isotopes of the original element. For example, it is possible to end up with Cl-36, or a chlorine molecule with nineteen neutrons rather than the nonradioactive eighteen or twenty of Cl-35 and Cl-37. You can also find Be-10, which is beryllium with ten protons instead of its typical four. Since nature strives to be in balance, these isotopes are not stable. They change, and these isotopes in particular are radioactive, which means they decay at a constant rate to reach a stable form. We can use this information to piece together events from the distant past.

The cosmic rays are mostly absorbed in the first 3 to 6 feet of surface material, so geologists can use cosmogenic nuclide dating to determine how long a surface has been exposed to cosmic rays, or how long it has been buried away from cosmic ray bombardment. In the context of the ancient lake in Panamint Valley, we were trying to date when the lake waters had receded and no longer covered the rocks we were sampling. When the waters were in place, the rocks would not have absorbed the cosmogenic nuclides. Once they receded, however, the surface rocks would absorb the rays and the isotopic decay would begin. To do this kind of dating we would be hunting for rocks with calcium

(Ca) or potassium (K), since those are the types that can produce the isotope Cl-36 when bombarded with cosmic rays. Cl-36 is useful for cosmogenic nuclide dating on surfaces thought to be between 60,000 and 1 million years old, which fit the hypothesis of the age of Panamint Lake.

Steve, Fred, and Roger had identified areas in Panamint that they suspected were ancient lakeshores. It was our job, as free student labor, to dig down to retrieve the rock samples for analysis back at the lab. Our time would be spent digging holes 6 feet deep into Panamint Valley's prehistoric past, peeling back the layers of thousands of years of Earth's history. It sounds captivating when explained that way. It was arguably less so in reality.

We trucked down to the southern reaches of Panamint, passing signs indicating Ballarat was to the east. Steve said we could visit Ballarat, a true ghost town, if we had enough time after collecting the samples. This sounded fun, so we collectively agreed to dig as quickly as possible. Steve laughed and reminded us that the weather was predicted to reach 107°F for the rest of our time there. I tried to look stoic while wondering what it would feel like to bake to death.

After we reached a point down an ever-narrowing dirt road where it seemed foolish to try to take the vans any farther, Steve told us to exit and grab our gear. We hauled in everything we would need for eight solid hours of work, including water, food, shovels, and sample bags for the rocks we would be carrying out.

Each sample was supposed to weigh about 2 ½ pounds, so we were instructed to eat our food and drink our water to lighten our packs for the rock-laden hike back to the vans.

In yet another newbie mistake, I had scoffed at my classmates' fancy water bladders and tried to save a few bucks by bringing my 1-liter Nalgene bottle. The hot temperatures back at Furnace Creek Ranch and Steve's story of the desiccated hiker had convinced me to purchase a 2-liter disposable water bottle for the Panamint portion of the trip. I have since learned that a good, sturdy water bladder can go a long way toward improving comfort and ease of hiking when doing fieldwork, though I still carry a backup water bottle in case I puncture the bladder.

As we trooped across the gentle slopes of the valley, I saw the heat shimmering off the rocks. Everything felt brittle, as if all moisture and life had long ago been sucked out of this place. After walking a short distance that felt like miles, Steve and Fred decided we had arrived. Roger was already waiting there, having surprised us students by scampering over the jagged terrain like a cross between a parkour expert and a mountain goat. He barely talked, looked like something out of an old-world coal mine, and could practically fly across ground that was giving 22-year-old college athlete me difficulty. Geologists truly were a breed apart.

Steve set us up with shovels and assigned digging locations, and our worlds became reduced to digging dozens of grave-sized holes in the middle of the desert in temperatures hot enough to make a person question the existence of ice.

I sweated. I wished for better gloves. I drank water, and then drank more. I sweated more. I regretted not packing more snacks. I tried to judge how much longer we would be out there by the angle of the sun and came to the (likely incorrect) conclusion that I could just crawl into my grave hole and lie in the shade and no one would notice my absence. The sweat kept coming. My designated site was down in a small depression, and I was supposed to be digging straight into the side of a low hill. I was making quite an attractive little burrow, and since the opening was about 3 feet wide I could have slipped right in to get out of the sun. Just as I was about to make this questionable judgment call, Fred appeared behind me.

"Hi Fred!" I saluted him, as energetically as I could muster.

He stayed silent, examining my work. I wiped my forehead with the hem of my T-shirt, which ended up soaked through. My feet slid, slick inside my sock liner–less boots. The world tasted of sweat. I was soft, and I knew it. I hoped Fred didn't.

"Can you dig a little faster?"

His words hit me like a slap to the face. How could he not see the effort I was making? I was a history major, damn it! Extreme-temperature hole digging was absent from my list of areas to strengthen for college graduation. And Fred had yet to even lift a shovel! Who did this scientist think he was to critique my digging?

Fortunately, Fred walked away without waiting for a response. Nothing I would have said would have been gracious or helpful for my future career in any way. I chugged more water, spitefully,

and then decided to take a break. I threw down the shovel and
gloves in front of the grave hole and wandered down the wash for
50 or so yards. I peered at the strange scrawny bushes and cadav-
erous grasses populating the wash. The stones were a mixture
of many rock types, but I only recognized a few from my limited
geology experience. I found one that sparked my curiosity, its
alternating folds of dark and light red containing tiny mysterious black mineral flecks. It went into my pocket, and I made a
mental note to ask Steve what it was.

As I turned back up the wash to resume my digging, something stark white caught my eye from among the tans, reds, and
browns of the rocks. I moved closer and bent down to examine
it. A miniature, perfect skull was waiting there, its teeth and delicate bones in their correct positions. It was bright white, about
1 inch long and half an inch across, and not yet starting to break
apart in that way old bones do. I picked it up, holding my breath
so as not to make a clumsy mistake and crush my new treasure. I
could tell it belonged to a rodent by the huge orbital sockets, long
incisors, and wide, flat molars. The lower jaw was missing, but
otherwise the foundation of a living creature was waiting for a
would-be scientist to stumble across it during a mind-bendingly
hot quest for something entirely different.

I pulled out some extra toilet paper and wrapped the skull
inside, planning to take it home. A sense of purpose overtook
me, and the thought of seven more days of digging seemed like
exactly what I needed. For the first time in my life, I recognized

that the experience itself was where the real learning happened. Adding to the body of scientific knowledge was just a perk along the way. For now, I was the beneficiary of an all-inclusive learning trip to the incredible Mojave Desert. Someday in the future, I thought, perhaps I would be paid to go to spectacularly inaccessible places and try to unlock the secrets of the universe.

▲

At the end of the trip, we had collected bags and bags of samples for the lab. I had managed to successfully conquer my naivete about the outdoors, despite not exactly loving the process. Our collective smelliness had reached remarkable levels, and a few of the students at the farthest end of the hippie spectrum had decided to avoid showering even after we returned to Furnace Creek Ranch; they had committed to flying back to Massachusetts wearing their two weeks of stinky glory proudly.

On one of our last hikes for fun, I found a moment to ask Steve about my mystery rock. I handed it over, expecting some great pronouncement about its identity, origin, or some other grand insight about its nature. Instead, he handed it back to me and said something that altered the course of my life.

"I don't know what it is. We'll take it back to campus and try to find out, though."

In my entire career in the humanities, I do not think I'd ever heard a professor admit to not knowing something. So much

of the humanities is open to interpretation, and so many of the
people I had studied with were so insecure that admitting any
ignorance would have been unconscionable. With just a few simple words, Steve showed me that it is perfectly acceptable to not
know the answers—but that ignorance came with an obligation
to search for the truth.

# Mauna Loa

## THE LONG MOUNTAIN

JUNE AND JULY 2008, HAWAI'I

I SPOTTED MY backpack, its blue fabric pristine as it slithered along the baggage carousel. I hurried over and hefted it onto the ground by one of the padded straps, checking to make sure none of the approximately thirty buckles and straps had been damaged in the cargo hold of the plane. Satisfied, I heaved it onto my available shoulder and began listing to the right. My laptop—safely inside my daypack on the other shoulder—was not nearly heavy enough to balance out a fully loaded 70-liter pack. Blinking, I shambled into the thick Hawaiian sunlight.

Outside the Kona airport, the land's tropical perfume even managed to smother the smell of jet engines and grease. The passenger pickup zone was small, so I figured I would be easy enough to spot. I dropped my bags to the sidewalk, inhaling as

I took in the thatched roofs of the baggage claim huts. I saw a group of twenty-somethings coming my way, talking animatedly and laughing. They recognized me and waved. Introductions followed, and soon I was sitting in a Chevy Tahoe with no front bumper and the most badass rock-crawling tires I'd ever seen. It looked like it was capable of off-road work, and I was hopeful we'd be doing plenty of that over the three months I would be spending at the USGS Hawaiian Volcano Observatory (HVO).

My greeting party consisted of Marco, Ashley, Helen, and Jenny. Marco, tanned and wearing a puka shell necklace, spoke with a pleasant Italian accent. He was working on his master's degree as part of the deformation group at HVO. Ashley, similarly bronzed with caramel-brown hair brushing her shoulders, displayed a welcoming grin. She was Canadian and was studying geophysics as an undergraduate. She was with the deformation crew as well. Helen had chin-length blonde hair, a ready laugh, and was naturally paler than Marco and Ashley. Her accent immediately revealed her British origin as she mentioned she was part of the gas group. Jenny, her compatriot, had long, light brown hair and a slightly more reserved demeanor than the other three. Her work was also with the deformation folks.

My mind whirled, absorbing as much as it could about my new colleagues. As Marco drove, they explained how excited they were to come pick me up from the Kona side of the island. HVO was almost on the exact opposite side of the island, near its eastern flank. Most volunteer researchers flew into Hilo, the

largest town on the east side. Since I was a nearly broke grad student responsible for my own travel costs, the frequent flyer miles
my dad had gifted me only took me to Kona. Fortunately, my
soon-to-be boss, Frank, had more pressing things to focus on so
he let the other volunteer researchers borrow his self-modified
government-issued truck to collect me.

We left the tourist-filled streets of Kona behind as we began
to traverse the island, heading for the route between the two
colossal volcanoes, Mauna Kea and Mauna Loa, that make up
the bulk of the Big Island. Marco told me we were on what was
known as the Saddle Road, and that it was preferable to avoid
this two-lane highway in the dark or during rain. Visibility was
limited as the road snaked around low hills and through forests
of tall, stick-like trees with tufts of dark green foliage. Many cars
in the oncoming lane were speeding, and it was plain to see this
road would not forgive lapses in concentration. About half an
hour into our drive, we leveled out and to the south I could see
the head of a volcano perched over a low ring of clouds encircling
its shoulders like a heavy scarf. I asked Marco to pull over so I
could take a picture and he obliged. I was looking at Hualālai
volcano, Ashley told me, and it was unusual to see the summit
due to the blanket of clouds it almost always wore.

I pulled my Nikon D50 camera out of my daypack and swung
the door open to get a better view. A present from my parents,
the camera was sturdy and familiar and had accompanied me on
a few short field expeditions and several camping trips over the

previous two years. I had resolved to bring it along as much as I could during my time working in Hawai'i. Its black metal body cool in my hands, I adjusted the settings and snapped a quick vertical portrait of Hualālai nestled in the afternoon haze.

Settling back into the truck, I queried the group about why a thin haze blurred the southern landscape but was absent in the north. Helen, the gas expert, explained that the haze was vog, volcanic smog, from Kīlauea. Water vapor, carbon dioxide, sulfur dioxide, hydrogen sulfide, and more were released from the summit eruption, from Pu'u 'Ō'ō, and from the place where lava cascaded into the ocean. The wind blew it around the south part of the island, and it could reach all the way to Kona.

Pu'u 'Ō'ō. The name struck me as unusual, even by Hawaiian standards. Ashley explained that Pu'u 'Ō'ō was a cinder cone on the East Rift Zone of Kīlauea that had been erupting since the eighties. She and the others had recently camped out on the fresh lava flow field close to it. I listened incredulously as Jenny gave an enraptured recounting of the trip. The group was tasked with taking measurements and recording observations of the flowing lava, and they had been dropped off by helicopter since the area isn't easily accessed by roads. I felt the fuzziness of airplane fatigue drop away, a new type of thrill taking up residence in my chest. Camping next to flowing lava? I was in the right place.

Marco steered the truck onto the shoulder and pointed straight ahead, saying the volcano's name. Even with an Italian accent, the words *Mauna Loa* merited reverence. He didn't have to tell me this was a prime photo opportunity. I hopped out again and knelt, focusing my lens. No vog marred the view, and the mountain sprawled before me. Even zoomed out all the way, the camera lens couldn't capture the volcano's full bulk. Enormous is an understatement. Two gentle slopes crested in the sky, a gradual rise in slope belying its nearly 14,000 feet.

Mauna Loa is an archetypal shield volcano, one of the main categories of volcano. Unlike the recognizable inverted V of a stratovolcano, the lengthy grade of a shield volcano gives no clue that it contains a magma chamber beneath its surface. The main Hawaiian Islands are all the tops of undersea shield volcanoes, the term originating from the image of a warrior's shield on its side.

Ahead of me, tall grasses danced through late-afternoon sun, a golden carpet unfurled at the feet of this leviathan of volcanoes. Dark lava flows branded its distant flanks, silent scars all that remained of eruptions past. Overhead, stray clouds contrasted with the brilliant blue of the sky and the muted dark of Mauna Loa's vastness. It didn't loom over the landscape it luxuriated in its expansive embrace of the horizon, its arms reaching across the whole of the island. In that moment, I understood Mauna Loa *is* the island of Hawai'i. Its sheer size makes it unique, the visible history written in lava offering so many clues for inquisitive scientists. Now, it was my job to help understand this colossus.

▲

My first two days at HVO were a whirlwind of introductions and
paperwork. I met my boss, Frank. He was rangy, and the rogue
white hairs in his dark curly hair and beard were the only real
clues to his age. Frank had warm brown skin and a smile that
started with a spark in his eyes and slowly connected to the corners of his mouth. He wore large dated glasses, a T-shirt, shorts,
and "slippahs," as sandals are known in Hawai'i. He moved in
a calm, purposeful way even while doing mundane tasks like
showing me how to find my intake paperwork on the government website. I liked him instantly, but my nerves were pinging
on high alert because I wanted him to like me, too.

The forms were familiar, but the online tutorial on helicopter
safety was a bit outside the realm of standard new-hire materials.
When I had spoken with Frank on the phone before he finalized
my offer, he mentioned helicopter rides were a very remote possibility since I had applied to be part of the Mauna Loa mapping
project. Helicopter time is very expensive due to the high costs of
fuel and maintenance, so the Kīlauea geology team had first dibs
on flight time. Unlikelihood notwithstanding, all HVO volunteer
researchers take the training—so I did, too. The gist of it was to
never, *ever*, walk toward the back of the helicopter if you wanted
to ensure your head and other appendages remained attached to
your body. It's a good life lesson with or without the additional
hazard of working on an active volcano.

As I completed my helicopter safety course, Frank cleared
his throat in my direction. We were sitting in his office, along

with another volunteer named David. We had been working to the tunes of the local Hawaiian radio station Frank preferred. Tinny ukulele notes wrapped around us, appropriately mellow to match Frank's placid demeanor. I looked across the small, chaotic mountain of geologic maps, miscellaneous pieces of lava, and scientific papers that dominated the large table separating the volunteer workspaces from his desk.

"We're going to the summit tomorrow."

Frank's tone was casual, as though he'd told me we needed more printer ink. I immediately aspired to be so relaxed about telling someone they were headed to the summit of the world's largest volcano—nearly 14,000 feet above sea level. While my brain whipped into a frenzy trying to process my good fortune, Frank relayed some information about taking aspirin and drinking water that night and the next morning. Back at the volunteer house that evening, I packed and repacked my daypack three times. Aspirin and water consumed, all I could do was will the hours to move faster.

▲

At 6 a.m., I eased the monster Tahoe to a stop in front of Frank's house in the town of Volcano. One of the numerous perks of being assigned to the Mauna Loa team was the privilege of driving the most hardcore truck in the USGS fleet. The other geologists and Frank were waiting. I handed the truck over to Frank

and climbed in the back. Quick introductions were made, and I
learned the visiting scientists were from other USGS observatories. I was impressed not because I knew their work personally,
but simply by the reputations of their home institutions.

We made a pit stop at Kandi's Drive Inn on Kawailani Street
to collect our bento box lunches. The smell of fried Spam greeted
us as soon as we hopped out of the truck, but food was the furthest
thing from my mind. I fretted about getting altitude sickness,
or not being able to hike fast enough, or saying something that
revealed how little I knew about volcanoes. These four men
were seasoned professionals, all working in a field I was hoping
to enter. My two quarters of graduate geology courses seemed
pathetically thin. I tried to be casual, hoping they wouldn't
notice my nerves, and was relieved when Frank started explaining the trees we passed on the Saddle Road were ʻōhiʻa lehua, an
endemic relative of myrtle that is one of the first species to col-
onize fresh lava flows after an eruption. The tree and its striking
red pom-pom flowers play a role in Hawaiian mythology, as the
volcano goddess Pele was said to have fallen in love with ʻŌhiʻa, a
handsome young man who loved a young woman named Lehua.
When Pele learned she couldn't have ʻŌhiʻa, she turned him into
a tree out of spite. Either the other gods took pity on Lehua and
turned her into the beautiful scarlet flower that graces the tree,
or Pele herself did it after feeling remorse for her actions. Frank's
native Hawaiian heritage and undergraduate degree in forestry
were already augmenting my education in unanticipated ways.

▲

Mauna Loa is 13,679 feet tall, and that is only what's visible above the ocean. An active volcano, it is the largest mountain in the world when measured from its true base on the seafloor. Its last eruption occurred in 1984 and made the news around the world. Visitors to the Big Island rarely make it up Mauna Loa, since the much more tourist-friendly Mauna Kea is festooned with astronomical observatories and a paved road and is currently dormant. Mauna Kea is about 120 feet taller than Mauna Loa, but Mauna Loa is more than double its volume.

We drove as far as we could manage along a twisty, bone-jarring road littered with lava—because the road itself was *made* of lava. Calling it a road is almost too generous a description. Four-wheel drive was necessary. After we'd reached the limits with the truck, we grabbed our packs and began the hike to Moku'āweoweo, the summit caldera. Its name means place of the fish, Frank explained as we trekked across the shifting rubble. I silently pondered that name choice, wondering but not wanting to interrupt the flow of Frank's words.

I struggled over the jagged lava, chunks of rust brown rock moving unexpectedly every time my boot landed on what looked solid enough. My strategy was to stay on the high points, like scrambling over rocks where I'd grown up in Colorado. But those sandstones were stable and jumping from prominence to prominence was easy. Mauna Loa's upper slopes were not proving as cooperative. After what must have been the fifth time I nearly broke an ankle, my cursing drew Frank's attention. He and the others were about 30 feet ahead of me, and all of them seemed capable of traversing the lava with no difficulty. Frank let me catch up and asked me if I wanted to know the secret to walking on lava. His expression was deadpan, which I was starting to recognize as his default. I responded with an emphatic yes, and he demonstrated how the jagged *a'a* lava is extremely unstable in the high points of each flow. As the lava is pummeled by wind and shaped by water, the chunks move to lower, more stable areas. In short, stick to the depressions if you value your ankles. As Frank and the others moved off, I used my new wisdom and bounded after them, pleased and better equipped for Mauna Loa's challenges.

▲

My heartbeat pounded in my ears, each staccato pulse of blood audible as my lungs ached, squeezing the scant oxygen through my body. I had always been an athlete and growing up at over a mile above sea level in the foothills of Colorado's Front Range had conditioned my body to function well with less oxygen. Still, I hadn't lived in Colorado for eight years, and I was a few years past my top competitive sports days. My quadriceps were burning, and I questioned the wisdom of bringing my heavy camera and a telephoto lens in my pack. The light wind gusting across the lava's surface sucked the moisture off my lips, chapping them almost on contact. While the hazardous glare of sunlight on fresh powder is well-known to anyone who has spent time in snowy climates, I had not anticipated the intensity of the sun bouncing off dark lava flows. The gunmetal gray seemed to reach toward me, every ragged edge accentuated by the thin atmosphere and dazzling sunlight. In contrast, the wan blue sky looked drained, muted with the effort of illuminating the landscape beneath.

Frank and the others disappeared after cresting the highest point in view. I trudged upward, worry lancing through my body in time with the glassy crunch of my boots on lava. My inability to keep up with the experienced geologists whispered a reedy challenge, pushing it from my subconscious mind into consciousness. Yes, Frank had accepted my application. Perhaps he was wrong to have done so. Perhaps every other researcher he had was a hardcore hiker. Perhaps, if I survived the day's work, he would relegate me to office duties, determining I was better suited to lab work than field science. I took another step and froze.

The summit of the world's largest volcano yawned in the crystalline air, stretching into the distance with no end in sight. A herd of soft clouds loitered overhead, casting indistinct shadows onto the naked rock below. Not a shred of life was visible, save for the scientists making their way down the steep slope ahead of me. No birdsong, no gentle swaying of grass in the wind, no distant dogs yelping. All was still on the roof of the world.

At the top of Mauna Loa is a caldera. The term comes from
the Spanish word for cooking pot, which is apt for something
found on volcanoes. Shaped like a flat-bottomed bowl with
steeply sloping sides, calderas are evidence of a volcano's violent
past. When a magma chamber empties its contents onto the surface, the chamber hollows out, lacking the support the molten
magma once provided. The rock on top stretches over a void, and
this sets the stage for a caldera collapse. This is one feature of
volcanoes that forms not due to eruptive activity, but rather from
subsidence. Calderas form like sinkholes, and caldera-forming
events are spectacular. Only a handful occurred during the
twentieth century, and the sheer size of Mauna Loa's summit
caldera means its collapse would have been epic. Measuring
nearly 4 miles long by 1 ½ miles across, the summit caldera,
Moku'āweoweo, dragged out a sense of wonder I did not know
existed within me.

The parts of the caldera floor touched by the sun winked shiny
shades of bright gray, completely lacking the reddish-brown
tones found throughout the lavas on the volcano's exposed outer
slopes. Mauna Loa produces mostly basalt lava, a rock rich in the
minerals iron and manganese. Both oxidize, or rust, when they
encounter oxygen. When basalt lavas are a few years or decades
old, there is typically very little oxidization present in the rocks.
The longer a piece of basalt is exposed to rain and wind, the more
likely that weathering and oxidization will be evident. Of course,
if the basalt has lower concentrations of iron and manganese,

oxidization will not be as dramatic as in rocks with higher amounts of these minerals. From where I stood on the caldera rim, the floor was blanketed in shiny lavas winking in various shades of gray and silver. Now I understood why the native Hawaiians named this place Moku'āweoweo. The shimmering of lava in the thin air looked iridescent, shifting like the scales of a newly caught fish. These were extremely fresh lavas—on the order of decades—formed practically yesterday considering Mauna Loa has rocks that are a million years old.

The walls of the caldera were nearly vertical in places, with horizontal bands of rock alternating between dark gray and lighter layers with a reddish tone. Each layer told the story of a summit eruption past, the volcano working to build itself one layer of lava at a time. The walls rose a few hundred feet above the caldera floor on average, indicating how deep the collapse was. The sea of metallic lava spread out almost as far as I could see, and I imagined the roiling hellscape of molten rock that once laid claim to this now-tranquil landscape. I inhaled deeply, thin air at the crown of the world's largest volcano flooding my lungs, before starting down the narrow lava trail after my colleagues. This was the roof of the world, perched at the top of a gargantuan volcano whose bulk stretched far beneath the waves, literal miles below my feet. I couldn't tell whether I was dizzy from the altitude or the enormity of standing atop one of the giants of creation.

▲

We gathered around another small protruding knob of lava, a slight bump on the relatively smooth lava flows of the caldera floor. Our goal was to identify intact structures of the 1984 summit eruption, and then to collect samples for cosmogenic nuclide dating, again using the isotope chlorine-36, like we had during my Panamint Valley ancient lakeshore research. Since we know when many of the flows on Mauna Loa occurred, that means an intact piece of flow will have received consistent cosmic ray bombardment since it cooled. By testing samples with known ages in the lab, we can see how accurate the technology is and adjust the process as needed to consistently produce the most accurate results.

I asked Frank for a few pointers about the intricacies of describing basalt for scientific purposes. At first glance, it looks like a plain gray or red-brown rock, sometimes with hole-like structures throughout. The holes, called vesicles, are where volcanic gases were trapped inside when the lava was still molten. Aside from the vesicles, basalt would be boring to the casual observer. But under a hand lens—a geologist's small magnifying glass like those used by jewelers—basalt has much more wisdom to impart. Tiny minerals reveal themselves, some smaller than the period at the end of this sentence, containing data about the conditions inside the magma chamber that birthed the rock. Their abundance or absence provides clues about eruptions millions of years before human life existed, like real-world time machines. Frank had spent more than two decades working

on Hawaiian volcanoes and was the world's leading expert on
Mauna Loa. Learning from him was the geological equivalent
of receiving a message from the Delphic oracle. I was energized,
and relieved that the actual work of finding, collecting, and doc-
umenting the samples was well within my capabilities.

Each sample took almost an hour to locate and collect,
and our packs grew heavier as the sun swept westward. We
approached one of our last sites in good spirits, and I had over-
come my awe of the visiting scientists enough to begin asking
about their careers. Steve, tall with a thick white moustache and
laughing eyes, related stories about his work on other volcanoes
as we hiked. He was based out of the USGS's Cascades Volcano
Observatory in Vancouver, Washington. He was telling me about
taking a helicopter to do research on Mount Rainier's summit
when we arrived at our next target, which was apparently a very
large hole in the caldera floor.

A few feet beyond the toes of our boots, the gray surface of
the lava was broken into large flattened chunks, as if a giant had
dropped a plate made of lava onto the caldera surface. Beyond
the crunched lava fragments, the caldera floor plunged downward into endless blackness. The hole was about 30 feet across,
the lavas lining its walls the color of old blood, frozen forever
in lazy drips, a testament to the fluidity of these flows. The
funnel-shaped void narrowed as it deepened. Frank, always the
cool customer, informed us that this was a vent from the 1984
eruption. As the others swung their backpacks down, preparing

to collect more samples, I paused. The sound of their conversation slipped away as the magnitude of Frank's statement washed over me. The ground beneath my feet was younger than I was. When this land formed, I was two years old.

Geologic time has a strange ability to distort an individual's sense of cosmic significance. Human lives top out at around 100 years, and we often take for granted that our planet had existed for eons before we were even a glint on the flagella of our eukaryotic ancestors. In the United States, most of us live on rock that has been in place for hundreds of thousands, if not millions, of years. Sure, we hear about landslides and floods reshaping the landscape, and some of us will even experience a natural disaster that reshapes the physical world we inhabit. Geologists are more comfortable than most people when thinking of the Earth's 4.54 billion years of history, since we converse in mind-boggling timescales with regularity. It takes effort, and some flexibility, to become comfortable with the practice. I had just completed my first semester of teaching undergraduate geology at CSULA and one of the challenges was finding effective ways of visualizing the extent of geologic time for my students.

If all of Earth's history is compressed into a single day, early hominids appear at 11:58 p.m., two minutes before modern humans, who show up as the clock is striking midnight. Even that analogy does not always convey how fleeting our presence has been when compared to the timespan of the forces required to open oceans, build mountain ranges that scrape the belly of the sky, and then tear it all down into individual grains of sand on a beach.

Now, I was confronted with the opposite dilemma. I was standing on rock-solid proof of our planet's life. Our world is still one of creation in addition to the destructive side we usually witness when geologic forces are in play. Mauna Loa, the world's largest volcano, is still active. It had erupted last within my lifetime, before the tragic end of the *Challenger* space shuttle and the world-changing fall of the Berlin Wall had etched themselves on our global consciousness. I was walking on earth younger than I was, and it reminded me of a true marriage of art and science, the famous lines of eighteenth-century geologist James Hutton: "The result, therefore, of our present enquiry is, that we find no vestige of a beginning, -no prospect of an end."

Frank snapped me out of my reverie, asking if I wanted to climb down inside the vent and have a look.

"Are you serious? Of course! Can someone take a picture?" I grinned and set my pack down.

Steve took the camera from me and down I climbed, awe and excitement hot in my veins as I braced my body against the narrowing funnel. The jagged edges of the glassy basalt threatened to rip any exposed skin. Avoiding the tiny daggers was nearly impossible, and I was grateful for my neon green suede gardening gloves. About 25 feet down into the vent, I reached a point where I could go no farther. All beneath me was black, and I stared, straining to see down into the beating heart of the

volcano. No luck, so I turned my face upward and smiled for
the camera.

▲

Volcanoes, when not actively erupting, tend to bury their secrets
deep—magma chamber deep. They often have intricate plumbing systems extending miles beneath the Earth's surface and
trying to glean information about what is happening underground requires scientists to employ a wide array of monitoring
techniques. Volcanologists frequently specialize in one or more
monitoring area, and then collaborate with other scientists
studying different aspects of the volcano to create an accurate
picture of the volcano's activity.

Volcanoes are complex. Much like when a human physician
examines a person, volcanologists take measurements from the
whole volcanic system. Measuring the composition, quantity,
and frequency of volcanic gas emissions lends insight into what
is brewing below the surface. Earthquakes on volcanoes indi-
cate magma moving around underground and breaking rocks,
and seismometers detect earthquake waves traveling through
the ground. When lava or ash are produced, collecting samples
allows scientists to examine the chemistry of the magma inside
the volcano. Some magmas are more explosive than others, and
the amount of crystals present can help determine the volcano's
explosive potential. This is critical information when a volcano is

actively erupting, but when a volcano is dormant scientists must look even more carefully for clues.

In the summer of 2008, Mauna Loa was not erupting. The twenty-four years since its last eruption in 1984 seemed unusually quiet for the giant volcano that had erupted thirty-three times since the 1850s. The challenge facing Frank, as the only geologist dedicated exclusively to studying Mauna Loa, was to understand what was going on deep below the volcano's currently peaceful exterior.

Deformation is one area of volcanology that is almost self-explanatory. Changes underground can be measured as deformation, or distortion of the volcano's surface. If magma shifts from one side of a magma chamber to another, the rock over the magma bulges upward, and the area where the magma used to be sags downward. An obvious bulge in the flank of a volcano was impossible to miss before the 1980 eruption of Washington's Mount St. Helens, when magma pushing upward caused the volcano's side to bulge outward almost 500 feet. In the case of Hawaiian volcanoes, their fluid and only minorly explosive eruptions are not usually accompanied by deformation that is visible to the naked eye. In recent years, volcanologists have started using radar technology called InSAR to measure changes of even a few millimeters on the surface of volcanoes. These data are collected from instruments mounted on satellites, but the technology is not as effective in areas of steep or heavily vegetated terrain.

Another ground-based way of monitoring volcano deformation relies on global positioning system (GPS) technology, which
is the technology used in smartphones to help with navigation.
In volcano research, two main types of GPS surveys are common:
campaign GPS and kinematic GPS (kGPS). Campaign GPS surveys
involve leaving a GPS receiver on a benchmark—a site that had
been carefully and precisely located before the survey—for several days or even weeks. It can record movement of less than a
millimeter per year in three dimensions when deployed with a
larger network of GPS receivers. The recorded speed and direction of ground movement can help reveal magma movement or
the location of faults. kGPS surveys are used to study areas when
covering a lot of ground is necessary. By using a known location
for a base station, where the position has been accurately measured and recorded, multiple kGPS receivers can be deployed to
record measurements relative to the base station. Repeating this
process over time reveals if the ground is moving relative to the
base station's known location.

Frank wanted to see if Mauna Loa was bulging near the
summit or down one of its flanks. The volcano typically erupted
either at the summit or along one of its two rift zones, so learning if it was bulging and where any bulge was occurring could
help Frank narrow down where the next eruption might occur.
Predicting eruptions isn't possible—it's the closest thing to
the holy grail of volcanology—so scientists are always working
to better understand eruption precursors. To that end, Frank

proposed deploying two teams to conduct kGPS surveys on Mauna Loa.

Our teams would drive to the summit in Frank's monster Tahoe, and Frank would bring our gear and the kGPS equipment via helicopter. From there, one group would survey Mauna Loa's summit and take the Tahoe back down when they finished. The other group would ride in the helicopter across the caldera and then hike the historic and difficult 'Ainapo Trail down the side of the volcano, camping out overnight. I was on the 'Ainapo crew with Frank and his other volunteer researcher, David. Ashley, Marco, and Kelly were taking the summit route. Kelly, warm and funny and very kind, had a bachelor's degree in geology and had been hired as a temporary contract worker at HVO. She usually worked with the Kīlauea geology group, but also worked with Frank when needed since she had been a volunteer researcher for him before starting her current position. We piled into the truck and cranked up Veruca Salt's song "Volcano Girls" for the long ride to the summit. I drove, always thrilled to push the limits of the rock-crawling tires.

Once we arrived near the summit, I gave the truck keys to Kelly and we all headed to meet Frank and the helicopter. When the bird touched down, he passed out the summit group's gear. David and I climbed aboard and we lifted off above the caldera. Marco, Ashley, and Kelly's figures receded as the craft gained altitude, struggling to achieve lift in the thin air. Our pilot, David Okita, is a legend in Hawai'i, having participated in multiple dangerous rescues using his flying skills. Frank had told me there was no one he trusted more to fly in the dicey high-altitude conditions of Mauna Loa's summit.

Flying across the massive caldera gave me a wildly different perspective of features I had seen on foot during the Cl-36 sampling trip. The caldera was even bigger from the air, and we crossed it on a rough diagonal, bound for the southeastern flank of the behemoth volcano. The flight took only a few minutes, while hiking would have taken the better part of a day. The helicopter's skids touched down and we unloaded our gear before Okita lifted off, waving farewell. The hypnotic beat of the rotors faded away, and we stood in the stillness for a moment, inhaling the crystal-sharp air. A ladybug landed on my hand, and I showed it to Frank. He told me they sometimes become caught in wind currents and land here, near the inhospitable summit. I set the insect down on a ragged chunk of lava, wondering if the tiny inadvertent explorer would make it back to more favorable conditions. We grabbed our packs, divvied up the kGPS gear, and set off down the deceptively gradual initial slope of the 'Ainapo Trail.

The kGPS consists of an 8-foot pole topped with a saucer-shaped disk and two retractable arms that extend from its midpoint to the ground, forming a tripod. Affixed near the midpoint is the display, which is about the size of a child's shoebox. A cable connects the disk up top to the display, and its surprisingly hefty batteries go into the display box. All told, the apparatus weighs 18 pounds and is an unwieldy hiking companion. Fully loaded with my camping gear, camera, clothes, food, some of the team's cooking supplies, and a few kGPS batteries, my pack weighed 43 pounds. When I was carrying the kGPS, my load increased to 61 pounds. I weighed 130 pounds with clothes on, so I was carrying 47 percent of my own body weight down from nearly 14,000 feet elevation. The 'Ainapo Trail would not be forgiving of mistakes, so I tried to move as cautiously as possible while keeping pace with Frank and David. Yet again, Frank's long legs devoured the trail, while David's much shorter stature seemed built to handle the rugged terrain. I was holding my own, though, and the views of Mauna Loa's slopes unfolding to meet the endless azure of the Pacific Ocean commanded admiration.

We watched the display on the kGPS unit closely as it tracked our progress and we looked for benchmarks along the trail. These benchmarks are a series of previously identified and recorded locations where we would take a measurement to compare with the survey done the year before. A change in elevation (positive or negative) could help determine if the volcano was inflating or bulging along the survey route. The benchmarks are a kilometer apart and are marked with a bit of spray paint to make them more visible amid the sea of black, brown, and reddish basalt rock. At each benchmark, we had to place the kGPS unit carefully and initiate the measurement. Each one required a solid minute for the unit to record our precise location, plus the set-up time needed to take the measurement in the first place. The first several benchmarks went by quickly, with the three of us trading

kGPS hauling duties between each stop. Neither Frank nor David
was prone to random chatter and we fell into a working rhythm
with minimal conversation.

The landscape we traversed was barren. Not a plant, weed,
flower, or a single tree marred the perfect flow of basalt tumbling
over itself, frozen forever on the slopes of the mountain that
birthed it. This part of the trail was covered with ropey pahoehoe
flows, a highly fluid type of lava that moves smoothly over the
terrain and creates nice broad, flat surfaces for walking. After
a while the lack of natural noise became acutely noticeable; the
only sounds my straining ears could find were my own exhala-
tions and the hollow crunch of boots on shelly lava. I embraced
the auditory and visual deprivation of the landscape and continued marching toward the next benchmark. The pleasure of
the otherworldly stillness came from knowing that this work was
important, that I was helping to add a small piece of information to the body of knowledge about this volcano, so much vaster
in space and time than any individual human. Understanding
this place could only work if we all added pieces to the puzzle,
building a picture one exertion at a time, decade by decade. I was
operating on geologic time now.

▲

By our fifth or sixth benchmark, fog enveloped us. It was
so thick that visibility was limited to only about 10 feet.

The ghostly quality of the lava flows was heightened by the restricted vision, and the fog was soon joined by a light drizzle. We dug rain jackets and pants from our packs while the kGPS did its measuring duties. Frank's wet-weather outfit consisted of camouflage cargo pants, a teal and blue lightweight waterproof rain jacket, oversized gray gloves, and a baby blue beanie pulled on over his neon orange visor. David's clothing was much more muted, with a selection of drab olives and browns topped with a floppy-brimmed sun hat. I had a gray and black men's snowboarding jacket, and a pair of waterproof black pants I managed to pull over my baggy cargo pants. As we prepared to carry the kGPS into the fog, I noticed the angle of the trail sloping down into gray nothingness. The low visibility meant we were virtually blind, and the hiker in front quickly disappeared into the mist. The sound of rain hitting rock added a kind of insulating ambience. I fought the urge to close my eyes since they were becoming increasingly useless in the gathering darkness of evening and rapidly intensifying fog.

▲

The fog was like a blanket that limited sound's ability to travel more than a few feet. Frank, leading the way, had to stop and wait for us to be within arm's length before any communication could happen. Every five or ten minutes, he stopped to make sure we were still behind him in the miasma. In between breaks,

I experienced periods of near total disassociation from what we were doing. I felt separated from reality, with no plants, animal noises, or visibility. My mind was on its own journey, imagining I had found a land bridge across the River Styx, and was picking my way down into the mythological underworld of Hades. Abruptly, my reverie was broken as David and Frank appeared, standing in the gray soup ahead of me, decidedly non-Hadean.

Frank removed his water-blotched glasses and wiped his eyes, suggesting we camp in a lava tube off the trail so that we wouldn't need to set up a tent and could use our camp stove in the shelter provided by the rock. I asked Frank about the wisdom in that plan, since a few days before he had told me about how native Hawaiians used lava tubes to bury the dead. On islands made of lava, the practice was born of practical considerations. Going inside lava tubes that haven't been specifically cleared for tourist visits is considered disrespectful to Hawaiian tradition and those who have died.

I was not generally superstitious, but I did try to be respectful of tradition. I'd also never been a fan of the supernatural. As a kid, a literal run-in with a cardboard Freddy Krueger cut-out in my local Blockbuster store gave me nightmares for over a week. Ghost stories told with appropriate suspense and flair by my father were fine, because reality was always clear. If he said a thump on the roof was a creature jumping onto the roof of our car, I could point at his semi-raised hand and accuse him of making the sound. If he was regaling me with the Jersey Devil's nightmarish escapades, I could rationalize that I lived in Colorado and certainly the devil had no jurisdiction in the Centennial State. It's no wonder science's quest to eliminate uncertainties about the world appealed to me. At that moment, however, I had one concern in mind about our proposed campsite.

Hawai'i had its own legends. The nightmarchers, or *huaka'i pō*, are the ghosts of long-dead Hawaiian warriors who return on certain nights to march—without touching the ground—from where they were buried to sacred places on the islands. Encountering the nightmarchers required a living person to go indoors and lie face down without looking. If this respectful action was performed, then the mortal might avoid a violent death. To my mind, it seemed like sleeping in a lava tube was knowingly disrespectful and almost asking the nightmarchers to show up. The dense fog was helping my imagination churn out worst-case scenarios, but Frank, always direct, reminded us that he was part native Hawaiian. He assured us his ancestors would have no qualms with our plan before ducking into the tube. David and I exchanged a shrug and followed our boss into the dark tunnel.

After a dinner of warmed-up Progresso minestrone, we nestled into our sleeping bags and listened to the rain. Sleeping on hard basalt is uncomfortable even under the best circumstances, and I could feel some fragments poking into my back and hips. I cursed my lack of foresight in borrowing a sleeping bag for my time in Hawai'i from my petite roommate, Amanda, who was several inches shorter than me. My shoulders wouldn't fit into the mummy-style bag no matter how much I squirmed and tried to make myself small. I used a wadded-up shirt as a pillow, my face, neck, and shoulders cold while the rest of me was stuffed into the weirdly tight sleeping bag.

Lava tubes form when active lava flows cool and solidify at the surface while lava still flows beneath, making a conduit for the molten rock. If the molten lava all flows out, a hollow tunnel-like structure is left behind underground. Sometimes, the roof of a tube will collapse partially or fully, exposing the tube to the surface. The tube we were camped in was about 3 ½ feet high and had a few bridges of intact rock across its width. I was under one of these narrow bridges, while Frank and David were under intact, solid parts of the roof. I thought I was being clever with my choice of sleeping spots because I had easier access to get out of the tube if needed, but as it rained intermittently throughout the night I learned how naive I had been.

When the pre-dawn light started to brighten the sky, I realized I had barely slept, and what little sleep I managed had been filled with restless nightmares. I withdrew my cramped limbs carefully from the tiny sleeping bag, only to catch sight of the word "child" on the faded tag. I exhaled through gritted teeth. Of course Amanda would fit in a child's sleeping bag. Of course I would absolutely not. Hurriedly, I shoved the bag back into its pouch, not wanting to divulge my poor preparation to my fellow geologists. My science skills were improving much faster than my camping ones. Groggy and stiff, I gingerly inserted my contact lenses with grimy fingers, adding only a little bit of dirt in the process. Blinking to clear my vision, I clambered out of the tube and was greeted with one of the most memorable views of my life.

From 9,000 feet up on the side of the world's largest volcano, I had a view to the east. A bank of clouds lay directly ahead of and below me, past where the lava flow we were on sloped out of sight. A few wizened 'ōhi'a lehua plants struggling to survive in the oxygen-poor air dotted the flow, softening the volcanic landscape. Behind the first bank of clouds, the sky opened, and the view stretched all the way to the coast. Rising out of a flat area of the landscape, far below my perch, was a continuous, roiling whitish-gray plume. Even from more than 10 miles away, the plume's convection was apparent. A twisting mass, the cloud it produced stretched to the southeast, wrapping all the way around the island as far as I could see. It was Kīlauea, famous for being the most active volcano in the world and the home of Pele, the goddess of fire and volcanoes. The summit caldera of Kīlauea has a few craters, and only three months before my arrival a new vent had erupted in the Halema'uma'u crater. While Kīlauea had been erupting continuously since 1983, the eruptions before March 2008 had remained on the flanks of the volcano. The most recent summit eruption to last more than one day was in 1959, when Kīlauea Iki, a pit crater next to the summit, erupted with lava fountains that reached as high as 1,900 feet.

Impressive though it was, the 1959 eruption only lasted 36 days; the 2008 Halemaʻumaʻu eruption had already lasted more than three times as long.

Entranced by the beauty of the deadly plume of toxic gases thrust up by Mauna Loa's much smaller next-door neighbor, I retrieved my camera and snapped a few pictures before breakfast. I was in awe of Mauna Loa, but I was hypnotized by Kīlauea. The plume seemed to have a tortured, rhythmic breath of its own, and I craved an understanding of what brought it to life.

## Field Journal

### JULY 6, 2008

Give me 13,000 feet above sea level. Give me back-breaking work and interminable hikes over wickedly sharp lava flows. Give me meticulous data collection. Give me the dirty, dusty, sweaty grunt work that no one else wants. To hell with the physical discomfort of being cold or hot, or thirsty or tired. I can go places few have ever gone or will go. I can take helicopter flights above active volcanoes. I can descend into volcanic vents. I can hike into the middle of the desert to garner valuable research material. I can take a machete to a jungle or use the stars to navigate at night. I can use my body and mind together to unfold the mysteries of our planet, or even of the universe. I can still be that astronaut of my child fantasies. I can still discover something that might change the course of science as we know it. I love geology because it is tangible proof that I can still do anything I want. I love geology because it gives back all of those little girl dreams I thought that I had to relinquish in order to be a so-called responsible adult. I love geology because in explaining life the way no other field does, it also offers me the opportunity to write a chapter in the never-ending story.

The three of us ate breakfast and continued down the trail. More kGPS measurements awaited as Kīlauea's plume beckoned us ever onward.

# 4

# Kīlauea

THE SPEWER

AUGUST AND SEPTEMBER 2008, HAWAI'I

MONDAY MORNINGS AT the Hawaiian Volcano Observatory, the thirty-odd staff members gathered inside a nondescript conference room. Fluorescent overhead lights and drab gray walls belied the spectacular visuals to come, courtesy of the room's projector and screen. The Monday meeting was for HVO's scientists to hear updates about ongoing or new activity on the volcanoes, scheduled research activities, and results of concluded studies. Visiting researchers from around the world used the time to detail their research at home or plans for their time in Hawai'i. Predictably, the gas group provided data on exactly what the volcano was belching out, the deformation group showed images of the Halema'uma'u crater with concentric rainbows and plots of squiggly blue lines indicating Kīlauea's ground

movements, the seismic researchers talked about any noteworthy earthquake activity, Frank (the one-man Mauna Loa group)
related important news from the giant volcano, and everyone
waited in anticipation for the Kīlauea geology group's report.
The rest of the data were always interesting but seeing imagery
of a volcano in active eruption was the uncontested grand finale
of the meeting.

Since March 2008, when the most recent summit eruption
began, the Kīlauea geology folks had been collecting reams of
data, including videos, photos, infrared images, and even infrasound recordings of the summit. The eruption continued on the
volcano's flanks too, meaning lava was routinely pouring into the
ocean and erupting over land, miles from the summit itself. The
geology group was headed by Tim and Matt—in their thirties,
they were two of the youngest HVO geologists. They were tasked
with keeping maps of the lava "breakouts" on land in the interest
of public safety, as well as with collecting as much data about
the lava flows as humanly possible. Since the sudden, destructive
dawn of the modern era of volcanology brought by the eruption
of Mount St. Helens in 1980, the USGS has invested heavily in
understanding the forces driving all types of volcanic eruptions.
The goal is to spare people and property from the inexorable
forces of a volcano. Worldwide, an estimated 500 million people
live near active volcanoes. Either we learn how to understand
volcanoes enough to live with them, or we learn how to live with
the destruction they cause.

Kīlauea, with its low-viscosity and mostly nonexplosive eruptions, was recognized as one of the world's finest natural laboratories for studying the mechanics of volcanoes more than one hundred years ago. Modern scientists have used technology and data collected from eruptions around the world to finesse their knowledge of key indicators of a volcano's destructive potential. At Kīlauea, everything from the lava's flow rate to the chemistry of the molten rock is measured, recorded, and calculated. Comparisons with previous Kīlauea eruptions were made to help the HVO scientists understand what the 2008 eruption's behavior might be, so that people in harm's way would have as much time as possible to evacuate.

In 1983, the volcano began erupting in an area known as the East Rift Zone, and within a few weeks lava flows had entered the Royal Gardens subdivision, destroying houses. The eruption began construction of the cinder cone now known as Puʻu Oʻo. A single vent became the focal point for the eruption, and lava spattering from this vent gradually built a sizable structure. Forty-four lava fountains contributed to the 835-foot cone over the course of three years.

After this cone-building lava fountain activity, Kīlauea's eruption moved north and east along the East Rift Zone. In late 1986, lava from the eruption first reached the ocean, burying part of the Chain of Craters Road in Hawaiʻi Volcanoes National Park. After a few more years of producing almost continuous flows, a lava tube channeled flows outside of the park and into the historic town of Kalapana. The volcano buried one hundred homes, a store, and a church. The lava flows that covered these buildings ranged from 50 to 80 feet thick. One historically important church, the Star of the Sea Painted Church, was moved to save it from destruction. In all, the Royal Gardens and Kalapana Gardens subdivisions and the towns of Kalapana, Kaimū, and Kaimū Bay were partially or completely buried by the eruption. From 1992 until mid-2007, the eruption continued within the park, in and around Puʻu Oʻo, and with flows that often reached the ocean, adding land to the island's coastline. No lava was present at Kīlauea's summit during the eruption until 2008, when everything changed overnight.

At 2:55 a.m. on March 19, 2008, seismometers recorded a series of events, with a major event at 2:58 a.m. This event created a new crater between 65 and 100 feet across along the eastern wall of the Halemaʻumaʻu crater, and debris ranging from the size of marbles to 3-foot-wide blocks was found on the rim of the crater, more than 230 feet above where the explosion occurred. No lava came out of the new vent that night, but gas rich in sulfur dioxide poured out. The vent glowed orange in the darkness, and when scientists from HVO drove to the east rim of Halemaʻumaʻu to investigate the damage caused by the explosion, the sound of rocks breaking down in the vent was obvious. It was fortunate that the explosion happened at night, because blocks as large as a foot across were ejected from the vent and onto a popular public viewing area. The scientists and park officials responded by closing the Halemaʻumaʻu crater to tourists, and setting up a livestreaming webcam so all vent activity would be recorded.

Kīlauea's summit was awake and making its explosive voice heard for the first time since 1924. It was a new era for the scientists who sought to unravel the mysteries of the volcano whose name means "spewing" or "much spreading" in Hawaiian. Just how much would the home of the fire goddess Pele spew forth during this eruption? Could the observatory's scientists learn enough from this unique natural laboratory to help protect lives and property during future eruptions? When I arrived at HVO in June 2008 those questions dangled in front of us, scientists new and experienced, urging us into the unknown.

▲

During a Monday morning meeting in early July, the lead deformation team geophysicist, Mike, informed HVO's staff that everyone able would need to help with a survey of Kīlauea—inside the caldera. Each year, the deformation group conducted a survey like the one I'd helped complete down the ʻAinapo Trail on Mauna Loa. This year, the survey of Kīlauea took on added importance due to the ongoing eruption. We had already adopted overnight monitoring of the vent at Halemaʻumaʻu, and everyone took turns on duty. HVO had a watchtower that rose above the single-story building housing the scientists' offices and laboratories. Since the day of the first summit explosion in March, cameras housed in the watchtower were trained on the vent. A computer with access to VALVE, the USGS's proprietary volcano monitoring software tool, sat in the watchtower and displayed constant streams of data from sensors deployed across the caldera. Signals from seismometers and tiltmeters ran across the screen in real time, and both infrared and natural video streams displayed as well. With a few clicks of the mouse, a watchful scientist could isolate and dig deeper into any data recorded.

A special category of seismic tremor that is known as a "harmonic tremor" occurs at actively erupting volcanoes. Produced by magma or gas moving underground, it is distinct from the kind of seismic signal associated with earthquakes. During the overnight monitoring, we expected to see harmonic tremors and signals of rock falls or small gas explosions from the vent. Low- to mid-frequency shallow earthquakes or high-frequency deep earthquakes could signal a more serious explosion; if any occurred above a predetermined magnitude we had to alert the Scientist-in-Charge of HVO.

In addition to the seismometers, tiltmeters deployed around Halemaʻumaʻu were another important tool for keeping an eye on the eruption. Originally designed for military uses, these sensitive devices measure changes in horizontal level on the order of millimeters. For more than a decade, HVO scientists had been using tiltmeters to understand movements of the magma reservoir beneath Kīlauea's summit. Mike's deformation group oversaw the tiltmeters and had collected exciting data that showed the

volcano was experiencing a cycle of deflation and inflation that
correlated with when pressure built up and was released by the
volcano via explosive eruptions at the summit or magma dis-
charge along the East Rift Zone. Recording the volcano's tilt gave
the scientist with overnight duty a window into whether anything
explosive might be building beneath the ground. I had already
done some unremarkable stints as overnight monitor and was
itching to help more with the active Kīlauea work despite my
assignment to the Mauna Loa mapping project.

   As soon as Mike announced the need for help with the summit leveling campaign, I shot an inquiring glance over to my boss,
Frank. He gave a barely perceptible nod of his head. I smiled and
found myself unable to focus on the rest of the meeting. The
leveling survey would take me into the caldera of an erupting
volcano—finally. A few months prior, I hadn't even had thoughts
about active volcanoes. Now all I wanted was to get as close to
Kīlauea's eruption as possible.

   I had fallen for the primal allure of a nonliving entity with
the diametrically opposed powers of creation and destruction on
a scale beyond the scope of human influence. During my days
as a history major, I was driven by the desire to understand the
origins of civilization. When I began studying geology, I was
motivated to learn the reasons behind Earth's unique features.
The instant I stood in front of Earth's ongoing process of birth-
ing itself, I had to know more. In high school English class, my
teacher had hung pairs of words on the walls to use as writing

prompts. My favorite pair was "beauty" and "brutality." In the raw womb of the volcano goddess, I had at long last found the marriage of those two words. Looking at the plume, dangerous and mesmerizing on its journey up from the inferno, I realized I was hooked on volcanoes. The more active, the better.

▲

The leveling campaign went smoothly. We worked nonstop for the better part of a week in teams of four or five to survey predetermined lines using equipment much like the kind regular surveyors use. If it wasn't for the need to use respirators in addition to our high-visibility vests and hard hats while working near the actively erupting part of the volcano, we could have been surveying almost anywhere. I reveled in the chance to work with some of the other scientists at the observatory and become closer with other volunteer researchers. The camaraderie and steady stream of jokes made the hours slip away. It never felt like work, but an opportunity to inhale knowledge.

The first small explosive event since April had occurred a few days before we began the leveling campaign, so we had to be extra vigilant near the vent. I took each chance offered to be close to the vent, even though our work wasn't focusing on it or the lava lake that could sometimes be seen sloshing around inside from the rim of the crater. I hadn't seen the lake yet, but I devoured the images and descriptions from the scientists who had. When the leveling campaign ended, I returned to work piecing together aerial images of Mauna Loa's lava flows in a GIS program. It was important information, but my mind kept returning to the churning lava less than a mile from where I sat in front of a computer, clicking away.

Matt, a newer HVO scientist, was close enough in age to most of the volunteer researchers that he often came over to the volunteer house to socialize. Matt was a funny, intelligent, tech-savvy guy who was helping blaze new trails in the science of volcano monitoring. A tall, lanky Eagle Scout who used his weekends off to complete his goal of hiking every trail in Hawai'i, Matt became one of my favorite people at HVO—it didn't hurt that we both had an affinity for the TV show *South Park*. I told him my desire to join him on some of his work on Kīlauea, and he said he'd be happy to have me along if Frank could spare me. The night after I secured Frank's permission, I barely slept in anticipation of joining Matt on the East Rift Zone's active lava flow field.

▲

I slammed the door of the truck and picked up my pack. Matt and I had made the hour-long drive to the edge of lava flow fields and were geared up to hike the rest of the way. Our objective for the day was to map some fresh lava flows that were burning a *kīpuka*, or forested area that had been surrounded by new lavas. We also wanted to take a sample of the new lava for geochemical analysis in the lab. Since these flows were erupting at some distance from the summit magma chamber, understanding their chemistry and the minerals they contained would help us add to our knowledge of Kīlauea's eruptions. Landing a helicopter on fresh lava flows is dangerous, so we had to reach our targets on foot.

We set off into the flow fields, the lava underfoot fresher than anything I had yet seen. Undulating glossy black coils of lava poured over the landscape. Matt told me the shiniest flows, the ones that looked wet on the surface, had erupted within the last week. Since they were dark black, the flows absorbed a large amount of heat from the sun that was beating down on us. We were close to the ocean, but the air hung heavy and still. As we hiked farther from the forested area where we'd parked, the lava swallowed everything in view. These were fluid pahoehoe flows, the lava forming intricate ropey designs interspersed with broad, smooth sections. Brittle crusts of almost pure volcanic glass lined some of the flow edges, with the surface of other sections already shattered into rough hexagons.

A thin rivulet of sweat ran down my upper lip and around the corner of my mouth. My rolled bandana was serving its purpose as a headband, stopping sweat from reaching my eyes, but there was nothing to be done about the rest of my face. I glanced down again at the lava crunching beneath my boots, wondering why the temperature seemed to have rapidly shot up another 20 degrees. A thin red line was visible in the crack between two big flat pillows of gleaming black lava. It stretched the length of the crack. Heat radiated up; my feet sweltered in the two pairs of socks I'd put on under my thick work boots to avoid blisters. I squinted down at the red line and pushed my sunglasses onto my head. This wasn't a discolored or oxidized part of the lava flow. This was a tiny window into *actual molten lava*. The flow I had almost walked on was still flowing underneath, the solid black crust on top newly formed and likely weak.

Matt was a few yards away, on a different part of the flow. I called him over, excited and flooded with adrenaline over narrowly avoiding plunging my boot into an active lava flow. He inspected the flow and said we would need to stay alert for others in the area. We turned and headed toward the ocean, trying to work our way across the flow field to locate a fully fluid active flow for sampling. As soon as we moved away from the tongue of the flow with the glowing red crack the temperature dropped perceptibly.

Before long, we noticed we were paralleling an area that was fresher than what we were walking on. The lavas looked as if they had only stopped flowing a few hours before, and Matt spotted something in the distance. He pointed it out to me, saying it was a skylight. I wanted to take a picture of it, so we carefully made our way as close as we could.

The skylight opened into the bowels of Earth. About 3 feet across and 1 foot high at the widest point, the skylight was an area where the roof of a lava tube had collapsed enough to provide a view into the tube itself. A cascade of lava was rushing toward the ocean in a torrent, and I could make out occasional chunks of solid lava within the molten rock flooding the tube. Here was the planet, pouring out the same elements contained in stardust, reshaping itself right in front of me. I was a voyeur, witnessing an ancient act too raw and powerful for humans to comprehend. Creation and destruction. Two halves of the same whole. The process would repeat again and again until our world's fiery core cooled and convected no more.

▲

Matt gave me a few minutes to marvel at this deadliest of rivers before redirecting us to our task of locating an active flow. Another half mile or so of carefully hiking across the flow field, avoiding the dicey newest flows and checking the handheld GPS units with the most recent flow maps, brought us to the edge of an oozing silver puddle that reminded me of a thicker version of the T-1000 Terminator in its liquid form.

I gawked at the active pahoehoe lava flow, and Matt smiled. He seemed to appreciate my newcomer's enthusiasm. What I saw was something that my limited experience told me had no right to exist. Rock was solid, it was firm, if occasionally brittle. It was dependable. Houses, fortresses, tools, and weapons can be built from it. Rock was static. It changed only slowly, only with great effort or great force. Yet here I was, a few feet from silver rock that was clearly alive in a way that only science or wizardry could explain. While Matt dug the sampling supplies out of his pack, I thought about what it must have been like to grow up in Hawai'i before modern science could explain volcanoes. Of course, a goddess would have been a logical explanation for ground that could give birth to itself, sometimes destroying lives in the process. What other than something divine could make the solid earth turn to liquid fire?

Nudging me out of my reverie, Matt handed me a full-face balaclava and silver gloves that were made for someone with much larger hands than mine. I donned the balaclava with care, leaving my sunglasses in place. Then I pulled on the gloves and assessed the heat-resistant silver. It matched the lava oozing in the background. Matt passed me a metal coffee can filled halfway with water and a rock hammer that I would use to pull off some of the molten rock.

I approached the flow guardedly. My goal was to get close enough to stick the pointed pick end of the hammer into one of the flow's toes. As I drew closer, the heat grew more intense than anything I'd ever felt. The flow I was targeting was in excess of 1800°F, which is nearly four times hotter than the highest setting on an oven. It seemed as if nature had hushed itself unbidden— except for my heartbeat, which was jackhammering in my ears. I paused, eyeballing potential targets and not wanting to get closer to that outrageous heat until I knew for certain where I would strike. I set the coffee can down behind me and decided on a nice fat lobe of lava about 6 feet away that was slowly blobbing toward my right foot. Faintly, I heard a tinkle that sounded like tiny pieces of glass being crunched ever so gently. The lava was making an almost musical sound as the new flow rolled over the older ground beneath. Between that and the radiating waves of heat that were hitting me full force it felt like a dream.

I couldn't take the heat much longer, so I clenched my teeth and stepped toward the flow, right arm extended with the hammer pick pointing down. Suddenly, my eyes felt like they were being sandblasted. At Matt's direction, I had kept my sunglasses on, so I tried blinking. The awful feeling remained, and I recognized my eyes were dehydrating. I needed to hurry, or my vision might end up more compromised than it already was, and one errant movement could result in serious burns. I took one last step, shielded my eyes with my gloved left hand enough to deflect some of the searing air, planted my right foot 10 inches from the flow, and stuck the pick into the living silvery glob.

Feeling no resistance, I pulled up slowly, straining against the heat to see what was happening on the end of the hammer. The lava followed the hammer's path, some of its sticky bulk attached to the pick with the rest fighting to stay part of the flow. The taffy from hell stretched vivid and red, the insubstantial silver crust broken by the hammer, the flow's dazzling scarlet insides exposed to the world. I kept pulling and freed a glob, the molten rock tendrils oozing back to the bulk of the flow. I pivoted, shaking the hammer to make the glob release its hold. It fell into the waiting coffee can and the water inside crackled to life, boiling instantly thanks to the scorching lava bleb I had dropped. Steam rose from the can as the sample was hyperquenched, solidifying it and preserving the information contained inside its primordial makeup. As soon as the boiling stopped, I picked up the can and rejoined Matt at a safe distance from the flow front. Relieved to be in cooler air and ecstatic about all things lava, I couldn't stop grinning. We packed up the sample and trekked off to map the lava flow that was currently burning an isolated island of green amid this sea of black.

▲

A little more than a week after my first trip to the flow fields, Matt and I returned to map several more lava breakouts. We hiked about 3 miles from our car, splitting up to make the mapping process faster. Each of us had a radio and a GPS in addition to our standard daypacks. My pack held water, snacks, sunscreen, wet-weather gear, my rock hammer, a notebook, and my camera. Matt took the more inland route, while I was closer to the coast. I enjoyed crossing the flows by myself, pausing to marvel at the plume created by lava flowing into the ocean. I stopped to double-check that my GPS's track logging feature was not yet turned on, spotted the new flow I was targeting, and headed for it.

HVO had joined the ranks of GPS devotees and found excellent use for the commercial handheld units in mapping lava flows. A geologist could turn on the feature that logs the unit's position every three seconds and walk around the edges of a flow at a moderate pace to produce a map of new flow boundaries that was accurate within 9 feet or so. It was inexpensive and allowed for boots-on-the-ground inspection of new activity. The geologist conducting the mapping needed to note any unusual characteristics of the flow, but aside from that it was a straightforward process.

I began mapping the flow after turning on the GPS track log and was walking steadily about 8 feet from the edge of the radiating flow when a terrible pain hit the right side of my back directly below my ribcage. Over the years I had played more than twenty-five sports at levels ranging from casual to varsity college. I've been thrown off horses and stepped on; I've had numerous broken bones, sprains, and concussions; and I had fully internalized my father's words to "play through the pain." This pain, however, did not even have the decency to let me stand up. I dropped to a knee, the knifing feeling taking my breath away. Thinking back to times when I'd had the wind knocked out of me, I tried holding my breath to see if that would stop the pain. It didn't, so I sat down on the lava. The sun was high overhead, and I could feel the heat searing through my pants, but there was no comparison to the agony hitting my right flank. Moving as little as possible, I eased my pack off and dug for my water.

Neither drinking nor trying different sitting positions changed the intensity of my pain.

By this point, I was sweating about as much as I had while working in the Mojave Desert. The lower to the ground I was, the hotter the air temperature thanks to the heat absorption of the lava. The air above the flows shimmered, but no oasis—mirage or real—was forthcoming. The pain would seem to tail off, and then suddenly roar back with equal or greater intensity. I tried lying down on the hellish black rock, but heat and lack of pain relief forced me back to a squat. The heat was too much to take sitting down, so I let my heavy boots do the work of insulating me from the worst of it.

After fifteen minutes, with the pain showing no signs of relenting, I made the call to Matt. Our radio connection was not good, but he understood that I was in serious trouble and unable to do much but writhe around in my uncomfortable squat. He told me he would check with HVO and see what our options were for getting me off the flow field and to a doctor. I'm not sure how long I waited for him to radio back, as everything had started to shimmer. The air, the lava flows, the ocean entry plume, my own hands—it all had an unreal quality. When Matt got back on the radio, he urged me to drink some water. I complied and listened as he explained my situation. He was a few miles away, and closer to our vehicle than he was to me. The area of the flow field where we were both working was too hazardous for a helicopter to come and evacuate me. Trying to remove me that way would endanger the aircraft and its pilot. He asked if I could walk, and I told him I'd try.

Using all the concentration I could muster, I forced the pain into a corner for long enough to heave myself to my feet. I wobbled and looked around, hoping for anything to use to stabilize myself. I was surrounded by nothing but a veritable desert of fresh lava, the most inhospitable environment I'd ever been in and yet somehow still on an island considered a vacation paradise. I commanded a foot to move and managed to stumble forward, grabbing my side, the pain even more pronounced now that I was attempting perambulation. Walking normally was not an option, so I hobbled slowly in what I hoped was the right direction. Normally my internal compass is accurate, but I was far beyond the realm of normal and quickly becoming delusional with pain.

It had taken me less than an hour to hike to the spot where I was stricken. As I shuffled toward where I thought we had left the truck, the small undulations in the flow field were torture. The slightest elevation changes forced me to pause, gather strength, and then labor through new searing jabs to my side and back. I recognized danger when I noticed I had stopped sweating; I tried to force down more water. It wasted no time in coming back up, and the effort of retching made me give up on drinking. My only goal was the truck. Shuffle after shuffle, I stared at the toes of my boots and willed them to keep moving. I found that if I grabbed the most painful area of my back and side with my right hand and squeezed as hard as I could I was able to ignore the internal agony for a few steps. The effect was short-lived, but enough to help me up some small inclines. Matt occasionally called me on the radio, asking for updates on my position. I had stashed the GPS in a pocket of my pack and could only tell him I was still walking. It seemed like I had walked for longer than it should have taken to reach Matt and the truck, and a look at my phone showed I had been on my agonizing trek for a few hours. My vision had become limited to my immediate surroundings, so I didn't notice when the trees at the edge of the flow field came into view.

Matt noticed me from where he had been waiting anxiously to intercept me. From my perspective, it was like he'd appeared out of the sky. His face, uncharacteristically sober, mirrored the first real concern I'd seen from him in the two months I'd known him. He took my pack and helped me over the last few hundred yards to the trees. Kelly, who had been working at the office that day, was waiting for us at the truck. When Matt made the call to HVO, she drove down to meet us since she had first-aid training. I don't remember how I got into the truck, but I remember sprawling out on the back bench while Kelly tapped her fingers on my lower right abdomen, trying to get a sense of whether I had appendicitis. The thought had never crossed my mind, which I chalked up to my near-delirious state. My reaction to Kelly's efforts didn't illuminate the situation, so I looked at the ceiling to avoid seeing the worry on her and Matt's faces.

When we arrived at the Hilo Medical Center, the only facility with an emergency room on our side of the island, the intake staffer handed me a cup for the routine urine test. I clutched the metal rail next to the toilet and tried not to miss the mark. When I saw the brick red and opaque contents of the cup I was convinced that I was staring at my imminent demise. Matt and Kelly had to return to HVO to use the flow data to update the public hazard maps, so they left before I was able to tell them about my certain doom. I probably would have asked them to leave anyway, though. I hate seeming weak in front of people I don't know well, and I hoped to consider them colleagues for decades to come.

The rest of my time in the hospital was fuzzy and punctuated with the expected needles, x-rays, disgusting drinks, and a CT scan. The next morning, they released me with a halfhearted and inconclusive diagnosis of something between a severe kidney infection and appendicitis. Neither of those explained the sudden, debilitating severe pain, but the doctor did not seem concerned with finding a concrete diagnosis. I was disappointed with the look of professionalism, but this was the only game in town. They seemed to think I would be ok, so off I went.

A day after some pain medication, industrial-strength antibiotics, and several bags of intravenous fluids, I was feeling ready to tackle volcanoes again. The doctor, however, had not agreed with that plan and restricted me to light duty for seven days. As soon as I got back to my computer, I looked up how likely I was to die from either potential condition, weighed my odds, and resolved to get back into the field as soon as possible. I knew I drank less than the recommended amount of water, but I didn't think I was that far away from the amount I should've consumed. At least this was a physical problem, with a physical solution. It seemed some of my youthful sense of invincibility was still clinging to me. Fortunately, geologists are a resilient bunch and Frank was more than willing to take me at my word when I said I was feeling reborn. And I promised to drink massive quantities of water to flush any infection out of my system before we hit the field again.

▲

After a few days of taking it easier than normal, I returned to more fieldwork on Mauna Loa and Kīlauea. I went on helicopter overflights to observe the latest flows from Kīlauea, helped with cosmogenic nuclide sampling to date pits dug by ancient Hawaiians, and conducted a gravity survey of the Nīnole Hills on Mauna Loa to help determine if a proto–Mauna Loa existed there before the current volcano. Still, I yearned to get close to the summit eruption on Kīlauea. Every day I spent in the office I would take my lunch break in the observation tower or—if the throngs of tourists weren't bad—out near the low stone wall that faces Halemaʻumaʻu and the new vent.

Next to HVO was the Jaggar Museum, which unlike HVO was open to the public. Inside the museum, excellent exhibits explained the science and history behind the volcanoes and the people who live on their slopes. Uniformed park rangers lingered outside in good weather, fielding questions like, "Does the vent go all the way to the center of the Earth?" (no) and "When is it going to erupt again?" (it *was* erupting right then). Sadly, when the long-lived Kīlauea eruption finally came to an end in the summer of 2018, the evacuation of magma from the chamber caused a collapse of the summit region. The Jaggar Museum and HVO's building next door were both rendered unsafe for occupancy as the ground beneath them changed yet again at the whim of the volcano goddess.

A main goal of the HVO scientists shared by volcanologists around the world is educating the public about volcanic hazards. If it wasn't for an extensive public outreach campaign during the lead-up to the 1991 eruption of Mount Pinatubo in the Philippines, more than five thousand people would most likely have died. Cooperative efforts between the USGS, the Philippine Institute of Volcanology and Seismology, and local authorities effectively evacuated local communities and the Clark Air Base. Knowledge sharing between volcano monitoring agencies and public safety organizations is critical in assessing emergency situations like that on Pinatubo, and educating the public about the threats that come with living on a volcanically active planet goes a long way toward building essential public trust in scientists.

Tim and Matt, the geologists in charge of the Kīlauea geology group, had decided that the cameras monitoring the Halemaʻumaʻu vent from the observatory watchtower in mid-2008 did not have a good enough view of what was happening inside the vent itself. The rim of the crater, more than 200 feet above the vent, was unstable and prone to collapse into the crater below. Capturing a rim collapse event on video was important for understanding the nature of the collapses. The vent had now been open for almost six months and was showing no signs of stopping. With knowledge that a lava lake was churning inside the vent, Tim and Matt concluded they needed a more continuous eye on the heart of the eruption. Installing a camera on the crater rim would accomplish that, but it would take some engineering to keep the camera safe and supplied with power.

As most of the volunteer researchers were back attending classes at their various universities, Tim asked me and David to join him on the installation mission. We loaded a truck with a sturdy tripod, the small video camera, a crush-resistant hard case about the size of a shoebox, a car battery, some cables, and a 3-by-1-foot solar panel. Our gas masks dangled around our necks, ready for the potentially lethal acid gases of the vog plume. We drove beneath and through the plume, its sulfur dioxide slightly stinging my eyes even with the truck's air-conditioning recirculating the interior air.

Once we reached the old public viewing area above the new vent, we parked and began ferrying the equipment the few hundred feet to Tim's chosen site. He picked a spot that had a direct view down into the vent, upwind of where the plume blew most days. Whenever it shifted it would obscure the camera's view, but having a good shot most of the time would have to suffice. Tim pulled out the heavy-duty tripod and set its points on the ground. I snapped a quick photo of him with the monster plume and a precarious chunk of the rim looking poised and ready to collapse in the background. After deploying the tripod, Tim walked over, a serious look on his face.

"Hey, do you hear those loud banging noises from the vent?" he asked.

"Yeah!" I responded, "They're so loud! It's amazing."

"If you hear a particularly loud one, get ready to run," Tim cautioned.

"Oh, right. Rocks can be ejected," I said, sobered.

He looked at me intently, as if taking stock of my state of mind.

"You do realize we could die, right?"

I considered this, looking from Tim to the vent and back. When I met his gaze, I had an answer.

"Yep!"

"Ok," he said, then pointed behind me. "Can you hand me those pliers?"

We resumed work, and once the camera was secure in its case, rigged to the battery and solar panel, and the tripod braced with piles of handy volcanic bombs we collected from the area, Tim offered to take a picture of me looking into the vent. I knelt, putting my elbow on my knee in a volcanic re-creation of Rodin's *The Thinker*. While Tim found the best angle, I watched as the plume swirled aside and a flash of red incandescence from the vent revealed itself. The now-familiar scent of rotten eggs and the noises emanating from the throat of the volcano sparked the part of my brain that had grown up consuming Greek and Roman mythology.

## Field Journal

### SEPTEMBER 3, 2008

The noises issuing from the vent were otherworldly. I now understand perfectly why ancient Greeks and Romans believed that Hephaestus or Vulcan, respectively, was hammering away inside of the volcanoes. It honestly sounds like someone is forging things in the traditional hammer-and-anvil way. The booms are loud, metallic, and frequent. Sometimes it sounds like metallic popcorn, and other times it sounds like the resonating, drawn-out intonation of a gong. It's not always noisy like this. In fact, everyone is remarking on how unusual the noises actually are. I feel privileged to have heard them.

▲

The next few weeks were an eruption of activity, literally and figuratively. I helped perform gas geochemistry research around the crater rim, and a helicopter overflight provided hard confirmation of a good-sized lava lake inside the summit vent. The gas team needed help doing some analysis near where we'd installed the camera, and the plume took a brief break while we were right over the vent that allowed us to see the lava lake for ourselves, closer than anyone else had managed. This time, Kīlauea alternated between clanging rock fall noises and loud gas-rushing sounds that resembled jet engines in their frequency and volume.

My time at HVO was ending, but my plans for returning to CSULA to take courses and teach were put on hold by an opportunity I couldn't refuse. A friend of Frank's, Mark Kurz of the Woods Hole Oceanographic Institution, had asked Frank to join him on a research cruise off the coast of Hawaiʻi. Mark was a geochemist who specialized in undersea volcanoes, and he needed a geologist to help catalog and process the samples of seafloor basalts he was studying for isotopic analysis. Most of the cruise was devoted to microbiologists studying iron-oxidizing microbes that thrived in volcanic environments—more extremophiles!—so as the odd man out Mark would need an extra pair of hands. Frank didn't want to go for a few reasons, and when he asked if I wanted him to recommend me in his stead, I thought he was joking. After three and a half months, I should have known Frank never joked about anything as important as this.

Some creative wrangling happened with my department chair back at CSULA, as well as some incredible kindness from my professors who offered to teach my undergraduate classes and let me make up the work I'd miss in my graduate classes. Not only had I survived my trials on the world's largest and most active volcanoes, I was now going to study an undersea volcano. Although my experience at HVO was ending, it seemed like my time researching volcanoes was only beginning. My biggest concern now was how I—a nonboater from a landlocked state—would handle almost a month at sea with twenty-nine microbiologists and geochemists.

# Lō'ihi Seamount

## THE TALL MOUNTAIN

SEPTEMBER AND OCTOBER 2008, HAWAIIAN WATERS

THE WORD "CRUISE" conjured up images of sun-kissed travelers lounging poolside with fruity adult beverages while their gleaming vessel skirted lush tropical coastlines. Reality is a harsh master, I remembered as I clomped up the metal gangway of the research vessel (R/V) *Thomas G. Thompson,* my home and office for the coming several weeks. The salt air in the Honolulu harbor replaced the ever-present odor of rotten eggs I'd grown accustomed to after four months of living on the rim of Kīlauea. Acidic gases had been exchanged for the faint aroma of oceanic decay that defines the border where earth meets water.

The ship was in a frenzy of preparations, and I caught a glimpse of the remote operated vehicle (ROV) *Jason* on the stern. This submersible was a celebrity in the undersea research world, since its first iteration was used to explore the famed wreckage of the *Titanic*, discovered after decades on the ocean floor. Its cheerful yellow-and-blue roof rested on a maze of metal gadgetry that looked like a shop class's fever dream. One of the submersible's robotic arms was visible, and my pulse quickened with the thrill of being so close to a piece of technology whose sole purpose was to illuminate the darkest, most inhospitable parts of our planet. This was true exploration, the stuff of Jules Verne novels and gritty retellings of expeditions to the poles.

Before that reverie could consume me, a man with a clipboard and efficiency spawned by a tight departure deadline asked my name and pointed me toward my berth on the boat's port side. I was assigned to room #35 and would be bunking with Maureen, a research assistant at the Woods Hole Oceanographic Institution. The room was higher out of the water than I would have preferred, as this positioning meant I would be more apt to feel the ship's motion. While I'd never been seasick on short boat trips, those didn't push any limits the way three weeks in the open ocean would.

I rapped my knuckles lightly on the room door, not wanting to startle Maureen if she was inside. No response, so I cracked the door. The room was spartan, with industrial gray metal bunk beds against the wall on the right side and a porthole window on the left. A small metal desk was attached to a wall and a metal shelving unit occupied the space above the desk. A door opposite the entrance opened into a tiny shared toilet and shower room. On the far wall of the bathroom was a door I assumed led to another cabin. The restroom door was latched open to prevent it from flapping around. I shoved my bags up against a wall and happily noted that Maureen had claimed the bottom bunk. Bunk beds were an unfulfilled childhood wish of mine, so the chance to have the top bunk in a research vessel seemed like an acceptable substitute.

After stashing my stuff, I headed out in search of Mark, the geochemist from Woods Hole who I was there to assist. I found him one deck below my quarters, along with his research assistant, Josh, and Junji, one of Mark's collaborators who was visiting Woods Hole from an institution in Japan. Unfamiliar lab equipment was strewn around Team Mark's portion of the ship's main onboard laboratory. Our main work surfaces were metal lab benches with wooden tops, and our group had access to a broad, deep sink as well. The drop ceiling hung low, almond-toned metal panels perforated for ventilation purposes. The weary yellow fluorescent overhead lights gave everything an air of 1970s futility, which stood in direct contrast to my absolute excitement.

Fighting to present a calm, professional demeanor, I shook hands with Josh and Junji. Mark explained that we needed to reposition our supply of nitrogen gas, so Josh and I went off in search of the large steel gas cylinders, which were nearly as tall as me. It took both of us to wrestle them over the 12-inch-high lip of the lab doorframe and across the lab to where we could shackle them to a pole. Since the lab's aft door opened directly onto one of the ship's open decks, my guess was that the raised doorframe kept water from flooding the lab during high seas. I was already enjoying the physicality of my first contribution to the team. My muscles were primed and ready for exertion after the conditioning of four months of volcano fieldwork.

After the nitrogen tanks were secure, Mark showed me one of our main pieces of equipment. He called it a glove box, and it looked like a combination between a newborn incubator and something Homer Simpson would use at work in his nuclear plant. Resting on the lab workbench, the trapezoid of thick plexiglass sprouted metal nozzles and plastic tubing, and had a small access door with a fat airtight seal on the box's left side. The most striking part of the box was the front, with its dual rubber rings almost wide enough to accommodate a basketball. Attached to the rings and jutting into the box were more black rubber rings, narrowing the farther into the box they went. At the end of the rings was a pair of sturdy seafoam green rubber gloves—hence the name "glove box." It was a sealed workspace where samples could be handled in a pristine atmosphere of whatever gas was required.

Smiling, Mark explained that the nitrogen atmosphere in the box would allow us to pull up samples of lava from the seafloor, preserving the unique gases trapped inside without exposing them to regular air. In order to get the samples from the bottom of the ocean into the glove box, he and Josh had designed and built custom sampling boxes that could be opened and closed by the *Jason* submersible's robotic arms, drained of water once they reached the surface, and placed inside the glove box's pure nitrogen atmosphere to be opened up and prepared for further study in the labs back at the institution. The sampling boxes were charcoal-colored, durable-looking plastic cylinders each with a 7-inch-tall, silver, T-shaped steel bar on top—for the robotic claws to be able to grip and twist—and a small plastic nozzle for draining the water.

Our preparations were interrupted by a shipwide announcement to muster in the long, low lab area for a safety briefing. The crew lugged in bright red neoprene survival suits that the entire science party had to put on as part of the training. They looked like triple-sized baggy wet suits, and were designed to keep us buoyant, dry, and visible in case we had to go overboard for an emergency. Nothing breaks the ice at a gathering of mostly strangers than seeing some of the world's most esteemed scientists gyrating wildly into outfits that look like the offspring of Santa Claus and a scuba diver. We giggled at each other, veterans of past research cruises and oceangoing newbies alike.

Our orientation concluded with a tour of the relevant areas of the ship and then we finished our preparations. I met Maureen, who had a wide smile and a vast amount of experience with boats from previous research cruises and the work she'd done restoring them with her husband. I resolved to bring her any onboard questions I came up with and made sure my bunk was situated.

The next day we set sail for the Lōʻihi Seamount, the only Hawaiian volcano I had yet to visit.

▲

The R/V *Thomas G. Thompson* was painted the same flat industrial gray as the bunk beds it carried, its boxy exterior confirming
its role as a utilitarian vessel built for function, not form. Its
274-foot length and 52-foot width sounded spacious until you
factored in forty scientists and crew members, the inevitable
inconveniences that arise when everything except your bunk is
shared, and no chance of a break for three solid weeks. When I
wasn't helping Mark, Josh, and Junji, I was exploring the ship. I
wanted to understand the rectangular maze of its decks, for my
own sanity and occasional need for personal space, and because I
would be responsible for a four-hour watch shift when the *Jason*
submersible was deployed on a dive. My shifts would be from
midnight until 4 a.m., so familiarity with the boat while sleep
deprived struck me as an important skill.

When in the water, *Jason* was down for about forty-eight
hours at a time. The submersible being remotely operated
meant that we didn't have to worry about the safety of any
crew members down in the deep. Sure, you're stuck reviewing
instrument and video data above the water, but each *Jason* dive
carried almost zero risk for the research team and its shipboard
pilots. The disadvantage—at least from the perspective of the

pilots and scientists—is that sleep cycles are often obliterated while on a cruise. My night owl tendencies lent themselves well to the late shift, so I was pleased with my assignment and tried my best not to show just how starstruck I was to be on the same ship as *Jason* in the first place. Since the *Nautilus*—the iconic oceangoing submarine in *20,000 Leagues Under the Sea*—was fiction, *Jason* was the real-life pinnacle of marine fame to my mind.

*Jason*'s place on board was on the level below my cabin on the portside stern, so the left rear deck, and if I jammed myself against the porthole in my room, I could watch the steady hum of activity that surrounded the blue and yellow metallic explorer at all times. It seemed like the *Jason* pilots from Woods Hole were tweaking and adjusting parts of the submersible's robotic armature throughout our journey from the island of Oʻahu to the ocean east of the Big Island. I found time at meals to ask one of the pilots, Will, a bearded man with hands worn by work and a ready sense of humor, for information about *Jason*, which was a more complicated system than I had realized.

The ship and pilots were above the water, while a physical cable connected the ship to *Jason*'s brain, known as ROV *Medea*. The two units remain separate, both boxy and lacking any aerodynamics, but overflowing with sensors and metal reinforcements. They needed to be kept separate in order to protect the cable, since violent surface action could cause the cable to detach. Having *Medea* midway was an extra layer of safety for the valuable submersible. This cable was no joke; it was fiber optic, reinforced, and 6 miles long. Our dives would be to a little over 3 miles down, so we'd have plenty of cable to do the job.

From there, *Medea* links to *Jason*, relaying commands down below and sending real-time video and data back up to the crew on the surface. *Jason* only moves at about 1 knot, which translates to 1.15 miles per hour, but the *Jason* crew must be careful to avoid moving *Jason* faster than *Medea*. Severing the connecting cables would mean the end of our research cruise, and possibly the loss of a $7 million submersible.

*Jason*'s control room aboard the ship was referred to as the "van." It was rectangular and cramped, with around twenty computer monitors hanging on the walls and occupying the available desk space. When *Jason* was down on a dive, three pilots were in the control van. One covered the main driving, and the others controlled *Medea* and kept the *Thompson* as still as possible while the submersibles were in the water. Two scientists were needed for the duration of each dive to monitor the video feeds for anything of scientific importance. Our cruise had an abundance of microbiologists, Mark and Junji as geochemists, and Josh as an engineer. I was the lone geologist present, so the previous four months I'd spent gaining expertise on Hawaiian lavas meant I became the default for questions regarding unusual rocks or other geologic features. I was scheduled into the late late shift in the control van with a microbiologist, which meant I could ask them questions about their work, and they could do the same with me. Once we'd made it to the ocean above Lōʻihi, the real work kicked into high gear.

▲

Using a submersible to investigate the ocean depths is a challenging, highly technical, risky operation. The sheer number of moving parts essential to each phase of a dive was staggering. First, a piece of equipment called a CTD rosette was lowered into the water column. Its job was to collect data about the conductivity (and therefore salinity), temperature, and depth of the water, and to collect bottles full of water samples on its journey to the bottom. The CTD rosette on board the *Thompson* was big—over 7 feet tall and 5 feet across, with a metal frame that supported a circular array of long, narrow, gray plastic sampling bottles. A crane was used to hoist the rosette over the starboard side of the ship and into the water.

After the CTD rosette was safely back on board, the *Jason* team deployed a series of acoustic transponders on the ocean floor. Then, the *Thompson* cruised over the transponders to form a network that *Jason* and *Medea* would use to navigate in the deep. While the network surveying took place, the science teams were making sure their equipment was ready for the dive. *Jason*'s role was acting as a substitute for a human researcher. Its robotic arms could retrieve samples, its video cameras and sensors record data, but it wasn't designed to haul much back up from the ocean floor. Therefore, we readied a piece of equipment called an elevator.

The elevator wasn't at all like the kind in department stores. For starters, it wasn't enclosed. It had a broad, square platform with metal legs at about knee height. It measured roughly 10 feet by 10 feet and was made up of hard plastic slats that allowed water to drain through. The middle of the platform was occupied by a vertical metal pillar, which rose about 14 feet from the ship's deck. Attached to the top of the sturdy pillar was a series of eight yellow hard plastic balls, each about 18 inches in diameter. These were floats, glass spheres encased in strong plastic that allowed the elevator to perform its essential function: rising to the surface after *Jason* loaded the platform with scientific samples. To that end, the science teams covered the platform with a variety of sample containers.

Team Mark had two different types of containers on the elevator. I laughed at the first type, since they were just standard plastic milk crates. The idea was that *Jason* would use its robotic claws to break chunks of lava the size of watermelons off intact flows on the seafloor and place them in the crates. These samples were needed to establish what happened when rocks are brought from depth, to serve as a sort of control for our real work, and would be fine exposed to our regular atmosphere. We just needed a way to keep them in place while the elevator rose back to the surface. Using a multimillion-dollar submersible to collect valuable research samples from one of the most inhospitable parts of the planet and then retrieving those samples with a crate available at Walmart for \$12? This was what I loved about science. The most effective solution was often the simplest.

The second type of sampling containers were the ones designed by Josh and Mark with the T-bar handle for *Jason's* robotic arms. Those were significantly more high-tech than the milk crates, since they needed to prevent the samples from exposure to atmosphere. Mark was interested in preserving gases trapped in the erupted lavas and exposing them to regular air would allow those target gases to escape. Fastening a smooth cylinder to the elevator platform slats was far more challenging than a milk crate, but fortunately Josh and Mark had worked out a solution on a previous cruise. The cylindrical sample boxes were held to the platform with large metal clamps that encircled the bottom of the boxes and bolted between the platform's slats. A few of those, a few milk crates, and a few microbiology sampling containers, and the elevator's available real estate was spoken for.

To combat the buoyancy of the elevator's floats, several heavy metal weights were attached to the elevator's vertical pillar. Those, along with the weight of the platform and sampling containers, were enough to send the elevator to the ocean bottom. When the containers on the elevator were full, *Jason* would use its robotic arms to remove the pillar weights and allow the buoyancy of the floats to pull the elevator back to the surface. Simple physics! The best part about the elevator system is that it allowed

*Jason* to collect samples throughout each dive, with the ship's two elevators ferrying samples back to the surface while the submersible remained below. The National Science Foundation was funding a large portion of the research on this cruise, and they had sent one of their program officers, Brian, along. I was curious about the cost of an undertaking like this one, with so many people and specialized equipment required for several weeks. Brian informed me that this cruise consumed about $100,000 each day.

I tried my best to look unsurprised, but I walked away with my mind churning. I knew that of the twenty-nine people in the science party, at least nine were students who—like me—were most likely not being paid for our work on board. We were gaining experience, and I was hopeful I would be able to do my master's thesis on some aspect of Lōʻihi's geology, but we weren't seeing any of that $100,000 in cash. The rest of our group consisted of either postdoctoral researchers, whose pay was often notoriously close to the federal poverty line, or scientists with tenure or other solid standing at research institutions. While some may earn in the low six figures, the glimpse into the budget for an expedition like this showed me that most of the money for the research went straight toward fixed costs, like fuel, food, and equipment. It was immediately clear that, just like the scientists at HVO, marine scientists were in this for the love of science rather than financial gain. I mulled this over as the preparation for sending *Jason* to the base of the undersea volcano continued.

▲

The Lōʻihi Seamount is the Hawaiian volcano most people have never heard of. It was still a little over 3,000 feet below the ocean's surface on its long slog to breach the sea level like its brethren next door, but it was taller from base to summit than Mount St. Helens was before its catastrophic 1980 eruption. It wasn't likely to clear sea level for at least 100,000 years. Just imagining the sheer size of a volcano beneath our ship boggled my mind. Land-based volcanoes were easier to comprehend with bulk you could point to, traverse, and hit with hammers as needed. Lōʻihi, however, lurked beneath the waves, with the endless expanse of blue surrounding the ship revealing no hint of its presence.

Lōʻihi represented an earlier stage in the evolution of the volcanoes that now form the Hawaiian Islands, the stage where their entire role was of creation, with none of the destruction we associate with land-based volcanic eruptions. Submarine volcanoes build and build, adding countless layers of rock from their vents, covertly growing an island beneath the water. Unlike famous submarine volcanoes like Kick 'em Jenny in the Caribbean, Lōʻihi's summit was still too far beneath the surface to disturb the visible waters above it. Occasionally, boaters report witnessing explosions of steam and ash, along with tsunamis generated by underwater volcanoes. Instead of making its volcanic presence known to observers, in 1940 Lōʻihi was first mapped on a bathymetric, or water depth, survey as a regular, inactive seamount, or underwater mountain. After it produced a series of earthquakes in 1952, scientists speculated that it might be an active volcano. In fact, Lōʻihi is the youngest Hawaiian volcano, at about 400,000 years old. It has been active numerous times since that 1952 earthquake swarm, with analysis of a swarm in 1978 providing scientists with the evidence needed to acknowledge Lōʻihi as the active volcano it is.

Similar to Kīlauea and Mauna Loa, Lōʻihi has a central summit caldera and two rift zones radiating away from its summit area. Inside the summit caldera are three craters known as Pele's Pit, West Pit, and East Pit. Eruptive activity in 1996 formed Pele's Pit, and future eruptions could occur there. The presence of rift zones indicates eruptions have occurred along the volcano's flanks, not just at the summit. Since lava types fascinate me, I started to wonder if we'd be able to see different types of lava at different locations on the volcano. I was confident in my understanding of lavas that erupt on land, but knowledge of undersea lavas in the scientific literature I reviewed was limited.

Mark and I talked about the purpose of his work on Lōʻihi. As a geochemist, he was primarily interested in studying the gases produced by the volcano and trapped in the volcanic glass that is created when hot lava contacts cold ocean water. By studying these gases, which come from deep within the Earth's interior, we could gain a better understanding of the origins of our planet, our oceans, and even our atmosphere. Specifically, Mark was most interested in analyzing the helium-3 found in Lōʻihi's lavas. Helium-3 is a rare isotope of helium only found in abundance at a few volcanoes on Earth, including Lō'ihi and others in Samoa, Iceland, and the Galapagos. From that, Mark inferred that the mantle beneath Earth's crust must be different in these locations. One hypothesis was the mantle plume—essentially a magma superhighway—that was feeding volcanism in these areas had a much deeper, more primitive source than volcanoes with shallower plumbing. Studying helium-3 at these places could provide insight into the origins of our planet, and perhaps even the universe itself.

I asked Mark why this research cruise was mainly for microbiologists, and he explained the link between the reduced iron consumed by the iron-oxidizing microbes the biologists were studying and Lō'ihi's heat and iron-rich lava flows. I was already familiar with the idea that volcanoes and the extreme organisms that survive on them are potential keys to understanding the origins of life on Earth, and this seemed like another way to gather pieces of information about the deep history of our planet. One research cruise would not answer the question of how life began on Earth, but it would help illuminate more pieces to the intricate, exquisite puzzle that is humanity's quest to understand how it came to be. I was just excited to be part of the detective team.

▲

*Jason*'s first dive was fifty-six hours long. Everything went smoothly and Mark was happy with the samples we obtained for the biology and geochemistry teams. He trained me on how to drain the water from the sampling containers, put them inside the nitrogen atmosphere in the glove box, and then extract them in preparation for geochemical analysis of the gases within. I delighted in the careful procedures and found working with the man-sized gloves attached to the glove box's access ports amusing, if a little annoying.

Junji had resorted to wearing scopolamine patches behind his ears to help with his seasickness. Fortunately, I hadn't experienced anything similar, but I did have the occasional headache when I spent too long fidgeting in the glove box or doing too much data entry on my computer. Some of the other scientists on the cruise were downright miserable, but the work was too important (and expensive) to sit on the sidelines. Seeing my colleagues power through was a testament to their dedication and it made an impact with me. My desire to do a good job for Mark was expressed in the level of detail I put into my rock descriptions. I drew on my recently earned knowledge of lavas to make sure each sample was characterized as completely as possible. If someone read my description without seeing a picture of the rock, I wanted them to understand exactly what that rock was like. Photos can be lost, but thorough field notes are the foundation of good, useable data—along with accurate data collection and sound analysis.

The next dive planned for *Jason* and *Medea* was to an area on the southern edge of Lōʻihi called Ula Nui. This second dive would be much deeper than the first, taking *Jason* more than 3 miles below the ocean's surface. The ship's cranes lowered *Jason*, *Medea*, and an elevator into the water. The descent would take *Jason* a few hours, so the onboard science continued while we waited. A few scientists finished what they could and decided to get in a little recreation. The ping-pong table in the center of the main lab drew small crowds watching scientists try to play despite the rolling swells pushing the ship. Several balls disappeared under pieces of lab equipment, never to be seen again.

I was working away at data entry when murmurs began to shoot through the lab. A biology team member came by my station and said there had been an accident with the elevator. A loud boom was reportedly heard below decks and the ship seemed to have rocked unusually in response. No one knew the cause of the noise yet, but the *Jason* crew reported the submersible and *Medea* were fine and still descending on schedule.

Mystified, the science teams speculated on possible reasons for the bang. Some of the more experienced ocean researchers guessed the elevator was the likely culprit, but we wouldn't know for sure until we could get *Jason*'s cameras down to the bottom to investigate. If something had happened to the elevator, it meant that all our sample boxes might be lost. The good news was that these were empty, since the elevator was on its way down. The bad news was that the science teams would be out some specialized and expensive sample collection tools.

A few hours later, *Jason* was on the bottom and following the
transponder signal from the elevator. We watched the live feed
from the deep in real time on the ship's monitors. I strained my
eyes against the absolute blackness of the ocean depths, seeing
nothing illuminated by the submersible's lights other than tiny
bits of flotsam, swirling in and out of the frame like motes of dust
in a sunbeam piercing a dark room. After several long minutes, a
somewhat familiar shape materialized out of the marine gloom.
The yellow plastic spheres of the elevator, normally so cheery,
were twisted into ghastly caricatures of their original state. The
thick steel armature holding the spheres in places was warped,
as if a giant had decided to reshape them into a bizarre modern
art piece. Our sample boxes were still attached to the platform,
but the platform's legs were misshapen. The elevator, one of
our primary science tools just a few hours before, was now little more than an odd shipwreck, destined to rust on the ocean
floor forever.

It was quickly surmised that a sphere must have failed on
the descent, imploding with enough force to shatter the others.
The reduction of the elevator's buoyancy and the weights it was
carrying likely caused it to essentially freefall through the water
column until its devastating crash on the old lavas waiting at the
bottom. At over 3 miles deep, the pressure exerted on the spheres
approaches 5 million pounds. Failure at these depths is absolute.
The chief scientist, Craig, and the other team leads met with
the *Jason* crew to discuss a potential salvage operation. After

discussing a risky plan that would connect the elevator to *Medea*, they decided to abandon the elevator as we simply didn't have the time or resources to retrieve it. The decision was made a little easier because there was a second elevator on board the *Thompson*, so we could still take samples with *Jason* and the expedition could continue. *Jason* was also able to rescue some of the more expensive chemistry instruments from the elevator before the dive went on. None of the scientists were overly upset by the loss of the elevator, since this was yet another example of Murphy's Law playing out on an expedition. It's all part of the cost of doing research in remote, hostile places. Without knowing why the elevator sphere failed, however, we had to cross our fingers and hope the second elevator could make the 3-plus-mile descent safely.

▲

I settled into my graveyard shifts monitoring the *Jason* dives, sleeping late unless Mark needed me to help with anything time-sensitive. After nearly a week on board, I was beginning to long for land again. Each day, I made time to stand alone on one of the decks facing the Big Island, visible to the west except when the vog from Kīlauea's lavas flowed into the ocean and shrouded the island in whitish-blue clouds. On clear days, I could see the three plumes emanating from Kīlauea at once. The ocean entry, Pu'u O'o, and the summit vent at Halema'uma'u were all producing gas, and I took the opportunity to reflect on

the four months I'd spent adding to our scientific understanding
of this incredible natural phenomenon.

I was enjoying my time on the ship, but I missed hiking over
the glossy sable expanse of fresh lava, hearing the crunch of my
work boots on new earth, often the only noise other than wind
to fill the stark volcanic landscape. While we could technically
touch the rock samples *Jason* loaded onto the elevator, it wasn't
the same as walking to an outcrop or a still-oozing lava flow and
collecting the sample myself. I spent my free time working out in
the ship's windowless gym, and when that became too oppressive,
I found a vacant staircase on the ship's deck and ran the stairs.
The ocean breeze carried no scent of the island 14 miles away,
but I knew from my time at HVO that this was some of the clean-
est air in the world. Filling my lungs with the ocean-scrubbed
oxygen while watching the hypnotic rhythm of the waves, white
froth capping sinuous blue, I was at peace if not at home.

About halfway into the cruise and nearing the end of one
of my shifts, everyone in the *Jason* control van was yawning
at the wall of flat panel monitors. We were watching the submersible cruise slowly along one of the deeper parts of the
volcano, recording video for the project I hoped would become
my master's thesis on undersea lava flows, when a bizarre creature briefly appeared on the feed from *Jason*'s main camera. It
swam languidly, a long brown body speckled with small white
dots, its spade shape undulating silently on screen for a few
seconds before vanishing back into the darkness. Seeing large

animals at more than 3 miles deep is unusual, since the pressure is extraordinary. Using the submersible as a reference, we concluded the fish was at least 3 feet long. It was the biggest living thing I'd seen *Jason* reveal since a few days prior when we came across a translucent octopus crawling around an outcrop of lava. The octopus was at a much shallower location, so this fish was now the frontrunner for the biggest organism we had seen at extreme depth.

Since my knowledge of fish was limited, I turned excitedly to the weary microbiologist seated to my left.

"What is that?!" I gasped, my own languor forgotten in the excitement of discovery.

He turned to look at me, blinking through heavy eyelids.

"I don't know. I don't *do* macrofauna." His eyes returned to the screens, staring vacantly at the endless twisting of the old basalts.

I would have been less surprised if he'd told me we'd just seen a unicorn or discovered the cure for cancer swimming alongside a volcano 16,000 feet below the surface of the ocean. A cacophony of thoughts started ricocheting through my brain.

*How does he not know what that fish was? He's been out here before!*

*I thought biologists working out here would know about all the animals in their research area.*

*Why isn't he even curious what it is?*

*Isn't our job as scientists to be curious about the world around us? Isn't that at the very heart of why we do this?*

*I just study rocks and I still want to know as much as I can about this place. What gives?*

I spent the last half hour of the shift sitting in silence that was probably only awkward for me. The *Jason* pilots were absorbed in driving the submersible, *Medea*, and the *Thompson*. The microbiologist busied himself reviewing data. I took the long way through the ship to get back to my cabin, my thoughts wild. It was extremely probable that we were the only people on the planet to have ever laid eyes on that fish. Not the entire species, of course, but that individual. That small experience alone was enough to validate my entire presence on that ship, since it was true exploration. In the age of concrete jungles and prepackaged media, how rare and special is it to see something no one has ever witnessed? For me, it was priceless. That feeling of awe was why I had such difficulty reconciling my colleague's absence of curiosity. To journey to the vast Pacific Ocean, to spend nearly a month on a research vessel away from family, friends, and work, to have dedicated years to learning and researching parts of the world most have no inkling of, only to see a creature so unexpected, utterly alien, and previously unseen by human eyes . . . and then to dismiss that as not worthy of interest?

I tossed in my bunk, the ship rocking from side to side in huge arcs, each wave sending me rolling from the wall to the edge and back again. If the bunk hadn't had a 4-inch metal lip, I would've rolled right onto the floor 5 feet below. The ship must

have been oriented perpendicular to the wave direction, something that happened when *Jason* was on a dive and the pilots
needed the ship in a fixed location. Sleep would only have been
possible if I managed to simultaneously brace my arms and legs
against the sides of the bunk and also relax enough to drift off.
The strong motion of my bed combined with my racing thoughts
on the nature of science itself kept me awake until after the first
rays of dawn had found the ship.

When I made my way down to the lab close to lunchtime,
I found another microbiologist to ask about the odd fish. I
explained that it was brown with white spots and was shaped
like a guitar, and the scientist I was questioning told me it was a
guitarfish. At first, I thought she was joking but she assured me
it was a type of ray. I busied my eyes and hands with processing
some of our volcanic glass samples in the glove box, but my mind
was interrogating my assumptions about science and the people
who pursue it. After a few hours of work, I had reached a conclusion: for as long as I chose to pursue science, I promised myself I
would position my curiosity at the forefront of everything I did.
Lifelong curiosity is what compelled Darwin, da Vinci, Newton,
Curie, Lovelace, Sagan, Goodall, Ride, Earle, and every scientist
or explorer who has followed their example. Staying within the
known guardrails of a single discipline or subdiscipline seemed
artificial and limiting, an affront to the innate drive to question
and explore that led me to science. My ability to ask questions
would be the fuel that sustained the fire for knowledge I felt

burning hot and tight in my chest, driving me to challenge my own perceptions and the world around me, always in search of answers.

▲

The cruise was starting to wind down, and our group had made good progress. Mark was pleased with the samples we collected, and I had identified enough interesting features on the South Rift Zone of Lōʻihi that we agreed I could easily focus my thesis research on the video footage and samples we had. There were about seven identified undersea lava flow morphologies, or forms, when we began our work, but I had seen enough variation between flow types that I thought I could make a case for breaking out the different flow forms into even more precise definitions. I also hypothesized that the angle of the volcano's downward slope exercised a significant amount of control on the physical form of the flow, and I would need to spend hundreds of hours reviewing *Jason*'s video footage and using location data to see if there was a correlation between the degree of slope and the flow forms. I was thrilled at the prospect of having Mark as one of my thesis advisors, since he was a fantastic scientist as well as someone I enjoyed working with. The loss of the elevator had been the only hiccup in an otherwise solid cruise.

We had two nights at sea remaining and I was diligently attending to samples in the glove box, thinking about the avalanche of work and personal business awaiting me on the mainland. My two favorite professors at CSULA had volunteered to cover the geology laboratory classes I was scheduled to teach, so I knew I would be returning to students with solid fundamentals. But trading fascinating daily work on active Hawaiian volcanoes for routine graduate student life punctuated by the occasional California earthquake was going to be a challenge.

I scooped some of the olivine crystals I'd pried from the basalt into a small metal cylinder that would serve as an airtight sample container, and then pulled my hands out of the glove box's gloves and rubber rings. On my way to the bathroom, I offered a greeting to Bill, one of the ship's data experts, who was red-faced and shiny after a date with the gym, and ran through a mental list of the procedures I would need to do to finish prepping the samples before I could finish entering data for the previous dive. I returned to the glove box and stuck my hands back in, resuming the delicate work. A noise that didn't fit broke my focus, the sound of something heavy yet soft hitting the ground. The computer lab was to my right, across the passageway that ran down the center of the ship. My view into the lab was at an angle, and I didn't see anything amiss right away. I heard a few sharp exclamations, and some gasps from people closer to the passage. I looked more closely and was jarred to see the identical soles of two running shoes, toes pointing to the sky and heels on the deck below.

The murmur of worried voices rose over the constant throb of the *Thompson*'s engines and air systems as a shipwide medical emergency call came over the loudspeaker, directing the responders to the computer lab. A crowd had formed, obscuring my view of the sneakers. Frustrated at being stuck in a crucial point of my work, I tried to hurry the sample preparation along to see if I could be of help. I caught flashes of people starting CPR on the downed person as the loudspeaker crackled again: all nonmedical personnel were needed to clear the back deck of the ship so a rescue helicopter could land.

Remembering the ruddiness of Bill's cheeks as he passed me, out of breath and sweat-drenched, I assumed it must be him. My fingers flew through the motions, heart thrumming in my ears a worried flutter. I finished the essential steps and yanked my hands out of the glove box, wiping them on my pants as I hurried for the back deck of the ship. I knew from my first-aid training that every minute mattered in cardiac emergencies and hoped that the quick response had been enough.

I ran into my cabinmate, Maureen, and we threw ourselves into the effort to clear the deck. In a low voice, she confirmed that our stricken colleague was Bill. I relayed how I'd just seen him leaving the gym not two minutes before I saw his shoes across the hall, and her expression mirrored what I imagined my own looked like. After twenty or so minutes rearranging the masses of equipment on the back deck, our group of nearly thirty received another loudspeaker message. This time, it informed us the helicopter would need to land on a deck at the front of the ship. We were instructed to clear it just as we had cleared the rear deck.

The gathering dusk was creeping across the water with a chill that pierced the thin T-shirt I wore, vog clouds drawing in around the Big Island, shielding it from view. My ears strained to hear the reassuring, familiar thud of helicopter blades beating their way through the air. I rushed to help clear the deck, grabbing cables and pulling whichever way I was directed, using all my strength and hoping our efforts would be enough to allow Bill to get to safety, but losing optimism with every passing second. We were only 14 miles off the coast, but I'd never felt more powerless to change fate.

After several more minutes of all-hands exertion on the forward deck, the scientists drifted into groups of two or three, talking in hushed tones, or wandered off solo to stare into the oncoming darkness and wait for the promised help to arrive. I climbed one deck up and sat down, leaning forward against a railing, my legs below the bottom rail and dangling over the edge of the deck, a perfect vantage point to watch the helicopter's arrival. My forehead pressed against the cold metal, I closed my eyes to focus on the sound I was certain was coming any minute. The moments ticked past, and I could hear voices crescendo on the deck below. I opened my eyes, and saw some scientists turn and embrace each other. I heard some start to cry.

I scrambled to my feet and hurried down the stairs to the lower deck, and overheard a scientist questioning one of the ship's crew.

"What do you mean it's not coming?" she demanded.

"I don't know why. They just said it's not coming." The crewman's voice broke as the words found their way out.

So there it was. I blinked back the hot tears that suddenly flooded my vision, my subconscious acknowledging the unspoken reality that Bill wasn't going to be saved. I made my way to an empty corner of the ship's rear fantail, the farthest part of the deck and opposite the side where *Jason* sat when it was on board. I saw the mechanisms that brought the submersible back to the surface working, reversing the process they had been mostly done with when the first emergency call went out. *Jason* was almost to the bottom for the last scheduled dive, and the *Thompson* couldn't set sail while it was in the water. Even if the ship's crew had known immediately that no helicopter rescue would come, it still would have taken another two hours to retrieve *Medea* and *Jason* from their full dive depths. We were stuck, anchored in place by one of the world's most valuable scientific instruments, hooked on the end of a line that stopped us from making an unscientific but desperately human run for the coast and the faintest chance of help for our friend and colleague Bill. I watched through tears as *Jason* emerged from the depths without even the valuable data it had been sent down to collect. It didn't matter. What mattered was that we were going home with one less person than we started the expedition with. In that sense, we had failed.

That night, we ate ice cream for dessert. No one wanted it, but no one wanted to waste it either, and the crew needed space in the freezer to store Bill's body while we headed for Honolulu at top speed. Everyone on board, crew and scientist alike, was nearly silent. Word spread that Bill had a wife and kids at home. Finding I could no longer eat, I walked around the ship aimlessly. The reality of doing remote scientific research was now clear. No one goes into work expecting to die, unless their work involves being shot at. Even then, the plan is to do everything possible to avoid injury or death. Even my work around active lava and erupting volcanoes had seemed like calculated risks. Other scientists did this work daily, and they had long careers. It seemed logical that taking precautions could ensure safety, but Bill's death at sea had shown me up close how field research means taking risks in situations without guaranteed outcomes. The ship's crew was well-trained, and Bill received CPR almost instantly after falling to the deck. Several people tried for a solid hour to get a pulse, taking turns to give him the best possible chance to be resuscitated. It just wasn't enough.

Upon our arrival in Honolulu, the medical examiner met the ship and kept us on board as part of the required process when someone has died at sea. I was grateful Mark kept us busy making sure our samples were ready for shipping back to his lab at Woods Hole. When we were allowed off the ship, my equilibrium rolled as my inner ear struggled to adjust to the solid concrete of the dock. I staggered around like a drunk person trying to act sober. Mark told me the effect would wane after a few minutes, and we said goodbye to the ship's crew and the scientists who

were off to other commitments. My excitement about continuing my research with Mark and heading home to my pets in Los
Angeles after five months away was minimal, since my thoughts
kept returning to Bill and the family who would never get to welcome him back.

Before I boarded my flight to the mainland the next day, I
inhaled the unique, now-comfortable fragrance of Hawai'i and
held it in my lungs as long as I could bear, promising myself that
this would be a "see you later" to the islands, not a "farewell."
The process of scientific field research was the challenge I had
been searching for my whole life, a path with enough twists and
obstacles—and very real dangers—to keep me coming back again
and again, testing myself and my understanding of the world
around me in the effort to discover more of the secrets held in
the Earth itself. It was plain that research was one part scientific
discovery and one part self-discovery. Working on active volcanoes took the difficulty level of regular life and turned it up to
eleven, and I knew, viscerally, that I belonged.

# Nevado Salkantay

## THE SAVAGE MOUNTAIN

JULY 2010, PERU

JAGGED TEETH CUT skyward, defying the low-hanging clouds to come closer, testing the fatal promise of the mountain peaks. The surprise of the first-class upgrade was yet to wane, and I was grateful for champagne that took the edge off my nerves. Sleep-deprived and Spanish-addled, my brain was reeling from ten plus hours of flying nearly due south compounded by a few hours of attempted sleep in the crowded Hostel Kokopelli in Lima's Miraflores district. The hostel patrons were apparently nocturnal, which I learned while doing my best to sleep through pounding, arrhythmic bass. The music was still playing when I woke up to catch a cab back to the airport for my 5:45 a.m. flight.

Despite my jet lag, the dramatic view from my window seat was mesmerizing enough that I imagined tasting the crisp, thin air on my tongue. The rugged Andes stood in dramatic relief against the pale ice of the sky, their sharp outlines establishing their geologic youth. The Andes were still forming around 10 million years ago, which in geologic time might as well be yesterday since geologists often work on the scale of billions of years. But my particular sense of geologic time was distorted from spending the better part of 2008 and 2009 working on Hawaiian volcanoes with rocks younger than I was, and in the case of flowing lava, not even solidified yet. To me, the Andes seemed ancient.

I swigged the last of my drink as the plane began its descent, and as the bubbles cleared the jet lag I realized why champagne brunches are so popular. The spacious first-class seat allowed me to contort for optimal mountain viewing. We flew toward the sun and the former capital of the Inca empire, Cusco. I was thrilled to have the opportunity to help my friend Joe, and I felt buoyed by my new master's degree and recent fieldwork on lava flows. But glacial moraines were a wildly different science and I worried I wouldn't be able to contribute much to Joe's research. Sure, I had studied moraines in my geology classes, but the closest thing in Southern California are mudflow deposits—about as close to Andean moraines as house cats are to lions. Would I be able to justify my presence? Had I become overly specialized after focusing so intensely on lava flows for my thesis work? Would my Spanish be helpful to the expedition? Perhaps most importantly, would I even be able to concentrate on the work at hand while standing in the valleys that were home for so many centuries to the Inca, a culture I found fascinating?

▲

Cusco was a sea of shiny red tile roofs nestled in a protective bowl of mountains. At first glance it looked every bit the Spanish colonial city. When the Spaniards tore apart the formidable Inca civilization, they built their legacy on the rubble of the Incas' masterwork. In 2010, the white stuccoed Spanish walls still sit on top of twelfth-century Inca foundations—huge dark boulders fitted together flawlessly and without a dollop of mortar. The craftsmanship of an empire of over sixteen million people who did not have beasts of burden, a written language, or even wheeled vehicles had fared better over the centuries than the handiwork of their European conquerors.

I walked the uneven cobbled streets, skimming my fingertips across the cool granite of the Inca foundations. Touching something built with incredible precision by people who were seen as inferior by the Spaniards was sobering. Disease and bad timing had cost the Inca everything; vast amounts of wealth fell into the hands of Spanish conquerors. As I made my way around seated women selling colorful knit caps and trinkets, I rolled the Quechua words around in my mouth, trying them out next to the Spanish I was already familiar with. *Qurikancha, Saksaywaman,*

*Soraypampa, Inti Raymi.* And *Machu Picchu,* undoubtedly the most famous words in the ancient language.

Joe's flight delays allowed me to begin absorbing the timeless atmosphere of Cusco on my own terms. I found our *hospedaje,* the Peruvian version of an intimate guesthouse, in the historic San Blas district on Calle Choquechaka. The street was narrow and steep, with cracks between the cobbles deep enough to ruin the high heels of anyone foolish enough to wear them. A rusty red truck that looked to be from the 1950s with the words "Pizza" and "Hamburguesas" painted on its side sat across the street from La Casa Elena. I made a mental note to sample Peruvian pizza if I saw anyone working the truck—for scientific purposes, of course.

The only identifying mark on La Casa Elena was a small brass plaque above a triple-wide lacquered wooden door. The guest-house's namesake and owner welcomed me warmly and showed me to a comfortable room off the open central courtyard. My Spanish had proven up to the basic tasks of catching a cab and introducing myself, so my excitement was winning out over nerves.

▲

Joe had previous experience in Peru as both tourist and scientist. While hiking the famous Inca Trail for fun, he noticed some moraines—loose piles of rock that were pushed into place by advancing glaciers during colder periods. As temperatures rise, the ice melts away, and the rubble of the moraines remains in silent testimony to a much colder past. Joe thought these Peruvian moraines looked to be about as old as some in Europe, which were formed during the "Little Ice Age" of 1300–1860. While not a true ice age, the LIA, as it is known, is an important area of study for climate change researchers. A link in the age of Peruvian tropical moraines and those well-studied moraines throughout Europe would provide evidence that cooler temperatures had affected more than just the North Atlantic region. Joe had already done enough research on the topic to gain prime scientific real estate on the cover of *Science* magazine in 2009, which is akin to an actor or musician appearing on the cover of *Rolling Stone*.

Now, he wanted to flesh out his hypothesis with studies of more moraines in different parts of the Andes, and the National Geographic Society was footing the bill. Our expedition plan called for a few days of acclimatization at Cusco's 11,000-plus-foot elevation—necessary preparation before tackling several weeks at extreme high altitude with no prospect for emergency evacuation. We were planning to trek over a 16,000-foot-high pass and camp on the shoulder of a 20,000-foot-plus peak for the duration of our trip. The few days in Cusco would be critical for a successful expedition. Altitude sickness can incapacitate anyone, regardless of age or physical fitness. Experienced mountaineers with years of preparation have succumbed to the effects of high altitudes, which include high-altitude pulmonary or

cerebral edema (HAPE or HACE). Even mild cases of altitude
sickness can render a person confused, groggy, nauseous, or
dizzy. Cognitive abilities are impaired too, and that makes for
questionable judgment calls at altitude. The amount of oxygen
available to the brain and body decreases along with the air pressure, even though the oxygen saturation of air remains stable up
to altitudes of 70,000 feet. Between 11,500 and 18,000 feet the
risk of hypoxemia, or severe lack of oxygen within body tissues,
becomes extreme as arterial saturation of oxygen falls below 90
percent. Our work would be entirely within this elevation range,
and the most effective way to reduce our risk was to acclimatize
in Cusco for at least two days before ascending any higher.

My previous high-altitude work—considered anything over
9,000 feet—had gone without incident, but I had been in the
United States. Medical assistance and a way out were always
an option. Helicopter rescue was not possible where Joe and I
would be working in the Andes, since no helicopter could make
the flight in such thin air. High-altitude helicopter rescues do
happen, but in well-traveled areas like Mount Everest, and only
when a pilot is willing to risk their own life in order to save someone from imminent death. Once we left the populated tourist
trek areas, we would be on our own.

I was restless to set out for the trail, but Joe had promised
that the sights of Cusco would be more than enough to keep us
occupied during our acclimatization period. With difficulty, I

turned off my geology brain and switched over to my historian mode. In the process, I became enchanted with Cusco.

▲

Our wanderings through the city revealed three obvious periods of history, one stacked on top of the next like layers of cake. The Inca world was the literal foundation for everything that had come since the Spanish conquered the mighty empire in 1572. The brilliance of Cusco's founders was everywhere. We visited the famed Qurikancha, the temple of the *Inti*, or sun. Its walls were once covered in gold, which the Spaniards eventually took as a ransom for the Incan ruler Atahualpa. Afterward, they demolished the Qurikancha and built the convent of Santo Domingo on top of its foundation.

I paused in the courtyard—or *Intipampa*—feeling the sun warm my skin and imagining the people who stood in that same spot more than five hundred years before. The sound of flowing water caught my attention, and I turned to inspect the Incan fountains, still functioning after all these centuries. The air of the Qurikancha hung heavy and incomplete. The people who had begun this place had not finished with it when their civilization was torn apart. I felt uncomfortable, almost as if I was walking over a grave disrespectfully. My historian side knows that we can only understand the mistakes of the past by keeping their memory alive in the present, but certain holy places carry the weight of their peoples' vision—and tragedy—forever.

▲

Packing was complicated. We planned to collect over forty rock samples, each weighing more than 2 pounds. That meant traveling light on the return journey would be impossible. Unlike backpacking for fun or adventure, on a full geologic research expedition you increase the load you must carry, rather than simply reducing weight along the journey by eating your food supplies. The physical labor of collecting rock samples leads to even more hard work on the trek out, so geologists must factor the end load into initial packing.

Old expedition photographs show teams of sled dogs ready to pull equipment hundreds or thousands of miles. Our expedition was shorter because we were looking to gather a great deal of information about a specific area rather than trying to cross a great distance to survey uncharted territory. We would be working within a region only a little more than 15 miles across, but the incredible relief of the terrain meant we would be covering much more distance vertically than horizontally. Factor in the elevation and potential severe weather, and this type of work becomes difficult even for young, fit people like Joe and me. This expedition would lead us into the heart of the Incan realm, and we would be only about 7 miles from Machu Picchu throughout the trip.

We might as well have been crossing the Pacific, based on Joe's description of our coming isolation. The odds of running into tourists were close to zero, and we expected to encounter maybe one family of alpaca herders at most.

One way scientific expeditions can benefit the region where the work is occurring is through using local resources whenever possible. Joe had asked the owners of Casa Elena for help in finding a cook and two wranglers for our trip. They had put Joe in contact with Gerardo, a local cook who spoke both Spanish and Quechua. This was critical, since the pack mules we rented were owned by Quechua speakers, rendering our Spanish skills about as useful as Russian with the wranglers.

Since Gerardo often worked for tourists hiking the famed Inca Trail, Joe assured me that he was familiar with vegetarians and ready to make sure I was well-fed and happy for the duration. This was a bonus, since most expedition veterans are prepared to handle special dietary needs themselves. I was just thrilled that I could dodge a diet of protein bars. They work in a pinch, but having warm meals is ideal when making extreme demands on your body, day after day, with less oxygen than what you have at home. There are no prizes for being the scientist most willing to deprive herself of luxuries, but the quality of work will suffer if the entire team is not able to perform at optimum levels.

▲

Joe is a perfectionist about gear and spent our final day packing and repacking. This trait was common in great scientists, so I did not begrudge his efforts. We had checked all the supplies a few times, but when he wanted to do another inspection I told him I was going to go for a walk near the *hospedaje* and take a few photographs. He thought that was a good idea, since I had helped with everything I could at that point and he mainly needed to put his mind at ease. I laced up my comfortable running shoes and set out.

I was angling to find a location with a view of the Plaza de Armas, Cusco's main square, which sits in a sort of topographic depression, nestled into a lower area surrounded by streets that angle up toward the hills ringing the city. I started up Calle Choqechaka and then spotted a narrow alley to my left. Since narrow alleys in unfamiliar places typically make for good detail-spotting, I ventured down the path. The pedestrians-only alley was named *Siete Culebras*, which means "seven snakes" in Spanish. There are seven streets in Cusco whose names begin with the number seven. Everything in Cusco seemed to have a reason for being, a trait which originated with the Inca. Siete Culebras was sadly lacking in snakes, but I continued on in my search for interesting parts of the city.

A few twists and turns later, I found the view I wanted. My Nikon's lens clicked a few times and I slung the strap back over my shoulder and across my chest, inhaling the thin air. Ahead of me sprawled the city's historic district, and beyond rose a

caramel-colored hill with "El Peru Glorioso" mowed in giant letters into the grasses on its flank. The side street I had wandered
onto was named *Resbalosa*, which translates to "slippery." It was
only wide enough for a small sedan, and made use of smooth,
rounded river rocks instead of flagstones, bricks, or cobbles. I
could see that the section of the street that went uphill, away
from the city center, grew incredibly steep. It appeared slippery
indeed. The route toward the heart of the city featured a school-
yard and school building to the left and the photo-worthy view I
had admired to the right. I set out toward the city, gripping the
camera in my left hand and walking faster. I had spent about
half an hour sightseeing and taking photos, and I felt guilty that
I was not offering Joe a hand despite his insistence that he had it
all under control.

Calle Resbalosa ended in a small plaza with a large fountain
in the middle. The plaza was the entrance to the school I was
walking next to, and I could see excited kids emerging into the
open space. They yelled and darted, impromptu games of tag
beginning and ending too fast to determine a winner. A couple
in their twenties lingered by the fountain, immersed in each other's gazes, struggling mightily to ignore the shrieks of liberated
ten-year-olds. I walked closer, crossing from the small boulder
field of Resbalosa street onto the smooth stone of the plaza. A
particularly zealous group of tag players brushed past me and
I looked around for more, not wanting to be kneecapped by
tiny bodies.

I never saw the sewer.

Even if I had been looking down, I probably would have stepped on it anyway. Over the years, I have set foot on countless sewers, manhole covers, and other small subterranean entrances typically capped with solid metal or grates. We all do, since common practice in most places with enough infrastructure to warrant underground plumbing dictates that all grates and covers are securely fastened closed or barricaded from contact with the public. Somewhere, a Peruvian utility worker sits oblivious to the havoc their carelessness wrought on both my body and research expedition, before I had even begun the risky part of the journey.

My white Nike Air Max cross-trainer hit the rectangular steel sewer cover and kept on going, forcing the cover down into the stuff it was supposed to keep out of sight—and far, far out of mind. One side of my body plunged down, the ground beneath my right foot suddenly exchanged for foul brown soup. My upper body heaved forward, still driven by the momentum of that last fateful step. My lizard brain kicked in, thoughts of self-preservation thrown aside in favor of my prized—and expensive—camera. My left arm shot up, camera held high and far from the small barrage splashing up from below. My left knee bent tight, and my left butt cheek met my heel. My right palm smacked the gray flagstones in front of me, arresting my forward motion.

I will always be grateful that the sewer I fell in was only about 1 foot long, 6 inches across, and several feet deep. Its size meant there was no way I could have fallen any farther than crotch-deep, which is what happened. The sewer was just too small to accommodate the rest of me. Had my entire body gone into a larger sewer, there would be no expedition story to tell.

The instant I reached maximum possible submersion depth, my coiled left leg unleashed every bit of force I could muster. I launched up and forward, a bizarre imitation of my days as a high school sprinter, the faces of the schoolkids shifting from wide-eyed wonder to unfettered amusement. The little bastards laughed. My struggles caught the attention of the lovebirds sitting on the edge of the fountain, their astonishment competing with a sad attempt at masking their own laughter. I stumbled forward and steadied, glad to be back on terra firma. I glanced down, assessing the damage. Until that moment, my innocence was intact, and the world was still a kind and merciful place. Then came the horror.

Liquid feces coated my right leg from hip to toe. My once-khaki pant leg was a nauseating gray-chocolate color, as if Hershey had added cremated remains to its Special Dark recipe. Horrified does not begin to describe my initial reaction, which quickly became double-horrified with a side of extra horror when the smell hit me. Realizing that my left hand was still faithfully discharging its duty and keeping the Nikon far overhead, I brought it protectively down to my chest. Oh god, the smell! Independent thought began to flee, reason being the enemy of gallons of human waste. My eyes were watering, and my stomach had started to clench in anger, the promise of imminent vomit strengthening with each passing second. It was then that I saw salvation in the fountain ahead of me.

I claim no pride in my next actions, but terrible events demanded terrible measures. I took one step with my clean and wonderful left leg, and then tried to do the same with the disgusting mess of my right leg. My knee refused to bend. At all. Rather than worrying about this new and alarming development, I dragged my semi-useless and suddenly painful right leg behind me in an exaggerated limp. The kids were howling with laughter at my less-than-elegant performance. To their credit, the paramours seemed to want to offer help. They exchanged worried glances and words I did not hear but settled for looking concerned. I reached the fountain's wall and heaved my dripping right leg over the low barrier. It took nearly all my remaining energy, since the adrenaline from the fall had started to recede and the shooting pain in my quadricep was making its presence known.

I had never knowingly defiled a public fountain before. Watching the clear blue water carry the sick gray-brown away gratified me on a sublime level. I closed my eyes and shuddered. Bodily excretions have never been my strong suit, which is just one of the many reasons I opted out of considering medical school. The moment of escapism ended all too quickly with more young voices, these even closer than before. I opened my eyes to glare at the pack of Catholic school students, who were enjoying a free show thanks to the *gringa pelirroja*. While there are plenty of redheaded white tourists who visit Cusco, I doubt most of them come for the in-person tours of the sewer system. At this point, however, any patience I had for being the subject of elementary kids' entertainment had evaporated.

"*¡CIERREN LA BOCA!*"

Surprisingly, they listened. This *gringa* spoke enough Spanish to remind them who was older, and they slunk off in small groups, chastened but still daring to sneak backward glances at me. I began readying to verbally eviscerate any Cusqueño cop who happened to disapprove of visitors bathing in public fountains. The young couple had disappeared, probably alarmed by my willingness to scream at young children in public.

I wiggled my right toes and found that they still functioned. The water, which moments before had felt heavenly, began to chill. The fountain was not quite deep enough for the water to reach the top of the poo-line on my khakis. I tried bending my knee to lower into the water a bit and discovered that bending it was not just tough, but impossible. My brain sent the signal to move, but my quad just would not or could not execute the command. With clenched teeth, I hoisted my leg out of the fountain and set off for Casa Elena.

*Step, drag-squelch. Step, drag-squelch. Step, drag-squelch.* The pattern continued as I made my way in the general direction of the guesthouse. Progress was slow, and my thigh was screaming with each drag-squelch. It occurred to me that departing the next day for one of the most grueling hikes of my life may not be possible. I replayed the sequence of events again in my mind, trying to identify the source of the pain in my leg. The fall was only a few feet, and my calf, ankle, and foot all felt fine. I must have scraped my outer quadricep along the edge of the sewer on the way down.

Once inside Casa Elena, I yelled to Joe. He was on the second floor of the building and could see me in the small atrium at the entrance.

"What's wrong?" he asked.

"I fell in a fucking sewer!" My anger and indignation were on full display.

Joe hurried down the stairs, face registering surprise as he took in my soaked pants and shoe. The smell reached him a second later, and he blinked a few times. I related an abridged version of the accident while struggling mightily to climb the short flight of stairs to our room. It is very difficult to climb stairs with one rigid knee, and I do not recommend trying it unless you have handrails or a banister. Joe trailed behind me, concerned and mildly repulsed. I asked him to apologize to Elena and her staff for my trail of squelchy poo footprints as I headed for the most important—and most clothed—shower of my life.

▲

That night Joe worried as I searched the internet for clues about why my knee wouldn't bend. Three parallel scrapes ran down my outer right quadricep, each about 1 inch wide and 6 inches long. They were technically open wounds, so I also searched exciting phrases like "wound infection sewer" and "disease sewer wound." Searching for "urgent care Cusco" had yielded nothing useful. Joe and I debated the likelihood that I would die of something resembling gangrene in the next few weeks. He was more inclined to be cautious, he told me, because he was responsible for me being in Peru. I told him I would tough it out, and that I did not think I was going to die from Peruvian sewer plague. My sense of responsibility was with the expedition, and I was desperate to ensure Joe could get his samples and make National Geographic, our highest-profile funder, happy. I was certain he could have gone on without me, but I didn't want to leave him with the burden of doing everything on his own.

I Skyped with my fiancé, Carlos, back in California. We had met just a year before, and recognized our shared curiosity and drive instantly. Not only had he hung around during the completion of my master's research, he had proposed the day after I defended my thesis. Fortunately, he had already grasped how unpredictable fieldwork could be and tried to help. He could search much faster with his U.S. broadband connection, and he confirmed that I likely had a quadriceps contusion—and I had apparently done nothing to correctly treat it. No mild knee bending for twenty minutes after the injury, and my scalding-hot shower was the opposite of the recommended icing. Crutches are recommended when pain is present, and complete rest is needed for the first three days after the injury. When untreated, quadriceps contusions with deep bleeding in the muscle can lead to myositis ossificans, a condition where the injured part of the muscle calcifies. This is a common injury for football players and other contact sport athletes and could take weeks to heal. Yikes.

I ended the call with Carlos, and Joe found some ice. I applied it for as long as I could bear the cold. To convince Joe that we should still head out for the expedition, I limped around the room in what I thought was a sprightly manner. He looked unconvinced. I asked him to reassess in the morning and told him it felt better already. Sleep came slowly, my leg throbbed, I was mad at the persistent pain. Angry that my little photo excursion might cost us the expedition, I resolved not to spend the next few weeks in Cusco as a tourist with a bum leg.

▲

We loaded our gear into a van early the next day. My ability to get up and down the stairs—convincingly pain-free, thanks to a huge effort at looking nonchalant—was enough to get Joe to agree to head to our first campsite. We were due to meet our wranglers and Gerardo at Soraypampa, a common departure point for Inca Trail hikers. On the drive we passed through cinderblock towns, the institutional gray knifing into the natural landscape. Glaring ads for phone services and soda contrasted with hand-lettered signs advising of locally made food for sale. Pockets of litter had collected on the sides of the road, colorful insults to Peru's stoic grandeur.

Also strewn on the roadsides—and occasionally in the middle of the road—were the saddest stray dogs I had ever seen. Matted hair, crusty watering eyes, and ribs for days. It broke my heart and made me want to cuddle my dog back home. It was also a reminder that so many of us who lead privileged lives in wealthy countries need when traveling: tourist areas are not reality. Expecting the San Blas district of Cusco to be representative of Peru is like expecting Disneyland to be representative of the United States. Every nation has its public face and its private sadness, and this part of my Peru expedition was my first in-person experience of that truth. Understanding the reality of life in places far from home is important from a humanitarian perspective, and knowledge of the difficulties people face in daily living can even save lives. In some countries, handing an underpaid public servant or police officer a bribe can mean their family eats, and your work continues.

As we left Cusco behind, the villages grew farther apart and the mountains seemed to stretch closer to the sky. The van gained elevation and the road twisted in front of us, sinuous, with inconsistent and inadequate-looking guardrails near some of the steepest precipices. Our driver was careless at best, but I swallowed my apprehension and stared at the mountains and valleys unraveling out the window as we passed. I ran through geology lessons in my mind, recalling sedimentary petrology lectures on different river types and their implications for the evolution of a landscape. The evidence in front of me confirmed what I had researched about the Andes—they were new in geologic terms and a good deal of water was at work. My historian brain kicked in again, and I imagined the Inca tending herds of alpaca and vicuña in the rich soil of the valleys below the road.

After a few hours, we passed a modern lodge resting in a broad river valley. The building would have been equally suited to Vail or Jackson Hole. Large rock walls were topped with a thatched roof, an attempt to marry traditional Peruvian building materials with today's creature comforts, like dual-paned windows and an outdoor hot tub. Joe explained that it was where some of the more affluent Inca Trail hikers stayed during their quest for Machu Picchu.

The trail was so popular and fragile that the Peruvian government strictly limited the number of hikers allowed each year, and all hikers needed to have a guide. Joe had done the Inca Trail years before in a stripped-down version, and said that there were other less popular trails nearby that were attracting a growing number of visitors because of the Inca Trail's issues with crowding. The pay-off of the trek was arriving at Machu Picchu at sunrise on the final day of the trip. The Inca were meticulous in building their structures in alignment with astronomical phenomena, and events like sunrise and sunset, noon, and the solstices and equinoxes were incorporated into the design of sites like Machu Picchu.

The original plan for our expedition allowed for time at the end to visit Machu Picchu and potentially Ollantaytambo, too. Just like the Inca Trail trekkers, though, we had to earn our reward. For us, the challenge was to survive a few weeks in the shadow of Nevado Salkantay—Quechua for Savage Mountain—at elevations close to double those of Machu Picchu. Throw in my debilitating leg injury and the potential for snowstorms in July (always a hazard in the Southern Hemisphere's winter months) and it was a respectable challenge. I was thrilled.

▲

Our wranglers and Gerardo met us at Soraypampa, which was not far from the tourist lodge. A dozen or so other groups were there, small tent encampments dotting the gently sloping field. Since I was mostly useless for setting up camp in my injured state, I limped around taking photos. The trick was trying to avoid the cantaloupe-sized rocks covered in mossy grass that carpeted every inch of the field. I would normally have only noticed those if I was looking to study them, but the uneven terrain made walking exponentially harder. Gerardo noticed my difficulty and asked what was wrong, and I explained the sewer accident as best I could in Spanish. He seemed a little surprised that the sewer had not been properly secured, and observed that it was good that it was me rather than one of the schoolchildren who had fallen in. I agreed out loud, but my pride had yet to fully recover from their laughter. Telling him how I had yelled at them to shut their mouths made him chuckle, so I hoped I had earned some points from Gerardo.

I managed to snap a few good shots of the sun setting on the mountains and then limped back to our group to have dinner. Joe's excellent preparation skills were proven once again by Gerardo's full stash of vegetarian-friendly options for me. In Quechua, he explained to the wranglers that I did not eat meat. Between that and my limp, whatever respect I had gained from Gerardo through my willingness to scream at random kids was negated by the wranglers' radiating skepticism. I resolved to show them I was good with horses and mules, and perhaps that would earn me a bit more esteem. Horse people often share a universal language and I hoped that would prove true here.

After dinner, Gerardo served up coca leaf tea. As the child of FBI agents, I was more acquainted with the penalties for doing drugs than the drugs themselves. It took Joe explaining outright that, yes, it was coca-as-in-cocaine tea for me to understand what I was drinking and why it wasn't available at Starbucks. He and Gerardo assured me it was legal and even helpful for staving off the physical effects of spending time at high altitude. I drank it and noticed that my mouth tingled slightly, but in a pleasant way. Joe told me he had coca leaves for chewing later in the trip, too.

▲

Morning arrived gradually. The knife-edge mountain slopes huddled around the small open plain of Soraypampa shielded us entirely from direct sunlight. I judged that it would be close to eleven o'clock before any of the sun's rays touched our camp, and the ambient glow of the gathering dawn tinged everything a frigid pink. Clouds of my breath balanced in the thin air for short moments before dissipating. July in the Southern Hemisphere is the deepest part of winter, and our elevation amplified the morning chill.

The smell of frying eggs and onions drew me to Gerardo and Joe, already huddled around a small portable stove. Having someone else make breakfast while on an expedition was a welcome novelty. The others slurped coffee, and I had another coca leaf tea in anticipation of the hours of climbing to come. Unlike most of my field geologist colleagues, I didn't drink coffee or beer. Along with my inability to grow the standard-issue geologist beard, my beverage preferences have inspired no small amount of teasing over the years.

Packing was quick, and I nabbed a few moments to test my bad leg. Flexion was still almost nonexistent, but I downed another four Advil and decided I would be able to make it with the assistance of a decent walking stick. Joe had reservations, so we discussed the possibility of me riding the pack horse with Gerardo. None of the body parts I needed for staying on a horse were affected. Gerardo, not himself a horseman, went to talk things over with the wranglers. I watched their animated conversation at a distance, torn between wanting him to convince them that I could go and simultaneously hoping he would return with news that I would have to stay behind.

My pain tolerance was high, but we were heading into some of the harshest terrain in the world. I was concerned I would slow down the group too much, or injure my leg more severely, or even somehow cause the expedition to fail outright. Joe, in his role as expedition leader, seemed comfortable with my assurances that I could maintain basic mobility. Since we had worked together under grueling conditions in Hawai'i, he was relying on that knowledge combined with my assessment of my own physical abilities, and we both looked to the wranglers for that extra bit of help that riding the pack horse could provide. Part of expedition management is knowing your teammates and their limitations, which can change on a minute-by-minute basis in wilderness settings. Gerardo strode back over to us, his face smooth and inscrutable.

"Sí, es possible montar el caballo."

I swallowed hard and looked toward the steep valley, the narrow trail hugging the right-hand side. To the left the earth plunged out of sight toward the river below. The mountains ahead reared skyward, daring us. It was go time.

▲

The mules, sturdy canvas bags full of camp and research gear lashed to their flanks, navigated a worn trail strewn with sporadic gray rocks. Some of these rocks were just pebbles, while others were small boulders. They had come to rest on our trail thanks to landslides, floods, and stream activity. The pack horse brought up the rear, his load intentionally light in anticipation of my eventual need to ride. The wranglers and Gerardo led the way, with Joe and me at the end of our group. Pain shot through my stiff leg with every dragged step, but years of dealing with sports injuries had given me the mental capacity to compartmentalize. I've played high-impact sports on sprains and with broken bones, and I've always taken the advice to "get back in the saddle" literally, even with cracked ribs. The pain-suppression techniques I employed were helped by the stunning views, but nothing could fix the agonizing knowledge that I was close to dead weight.

To our left, the ground dropped away to a small flowing river, its gentle undulations a textbook example of how streams carve valleys over millennia. The riverbanks were flush with rich green vegetation. Joe and I discussed taking sediment cores near the river to reconstruct the ancient climate. Knowing the Earth goes through periods of warming and cooling, the scientific community was still gathering evidence about the details of these periods. Cores from this Peruvian river valley could provide valuable information about climate fluctuations in this region, and we could compare the cores to other data from locations around the world to paint a picture of the planet's climate throughout different eras. Cores could reveal the remains of prehistoric animals and plants, records of floods and landslides, volcanic eruptions, meteor impacts, and when it was dry or wet. This type of work was yet another piece of the paleoclimate jigsaw puzzle we were attempting to solve with our current expedition, but it would have to wait. We had no coring equipment and our agenda was tight enough as it was.

A gratifying aspect of geology for me was the "superpower" of understanding how mountains, rivers, and volcanoes all came to be. The forces that shape the world around us are tangible and still happening every day. Ours is a living planet, and the river flowing in the valley below our trail was sculpting the rock it flowed over, eddy by eddy, moment by moment. This aquatic force eroded the seemingly immovable rock, carrying microscopic pieces downstream, forever altering the world in a way too subtle for humans to perceive. The simplest geologic process can have mind-blowing results over time, as in my favorite example—the Grand Canyon.

I rode for twenty minutes or so to rest my leg. Despite Gerardo's interventions, I was unable to convince the wranglers that I knew what I was doing on horseback. Humiliation burned as I was led like a tourist on a guided trip. I didn't even have the reins to hold, so I shoved my gloved hands deep in my jacket pockets, petulant and resentful of my aching right leg. The horse ignored my quiet encouragement, focused on gulping the thin air and trudging forward.

We gained elevation steadily but slowly until we reached a slight right curve in the trail. To our left, the most vivid aquamarine oval flashed bright in the sun, glacial waters seemingly from another planet. The brilliance of the waters contrasted with the harsh white and gray of the looming mountain ahead of us, Nevado Salkantay, which Europeans first summitted in 1952. The Inca who made this landscape their home for centuries understood the weight of this place. The trail we followed was one of theirs, and it skirted the southern flank of Salkantay, gaining as little elevation as possible while steering clear of the large rubble field at the base of the mountain.

Salkantay is covered in glaciers, the ice thick and timeless, obscuring the mountain's rocky heart. This is the highest peak in the Vilcabamba range, and it transcends the surrounding landscape by a few thousand feet. For the first time, I understood the simplicity of George Mallory's famous rationale for climbing Everest, "because it's there." The stark white slopes of the mountain issue a challenge to all who pass beneath its massive shadow. Fortunately for the sanctity of Salkantay and the health of my leg, we were not there to answer the challenge of its existence. After we had rested and snacked, we angled to the east and resumed our trek up the steepest part of the trail. Inka Chiriasqua Pass waited.

Before we set out that morning, Joe had given me an overview of our route. Our path snaked over Salkantay's flanks, balancing between the Scylla and Charybdis of heart-stopping cliffs and ancient glaciers. We needed to summit one of the highest passes in the Andes to get to our field site, a difficult climb even when everyone in the party was fit and healthy. Gerardo told me that this section of trail was too steep for the horse to carry a rider, so I slid to the ground, girding myself for the ascent.

The Quechua name *Inka Chiriasqua* translates to "the place where the Inca rest." At over 16,000 feet in elevation, it is higher than every mountain in the continental United States, and tops all but four mountain peaks in Alaska. The trail leading up to the pass runs almost due east, traveling roughly parallel to Salkantay's south face. The ethereal glacial lake sits to the left of the trail, ringed by dark rock, reflecting the brilliance of the sun and snow, giving the landscape the hard edge of a cut diamond. Perhaps counterintuitively, the path itself does not follow natural depressions. It tends toward the right of the lowest possible route, higher up the slope facing Salkantay's south side. The path is about 2 feet wide, with little margin for error due to the precipitous drop on one side and the unforgiving rocky incline on the other.

Initially, I was annoyed at the extra effort required to gain elevation and stay upright. My limp made my gait irregular, my left foot striking the ground faster and harder to compensate for my weakened quadricep. I glanced ahead toward Inka Chiriasqua—still half a mile and over 500 vertical feet distant—pulling my focus from the uneven path. My right leg *draaaagged*, followed by the *crunch* when my left boot slapped the jagged pebbles.

*Draaaag-crunch, draaaag-crunch, draaaag-crunch, draaaag-cru...*

"Aaah!" I cried in surprise.

The softball-sized rock beneath my foot tumbled down the slope to my left, cracking into boulders the size of Great Danes and Honda Civics, gaining speed as it hurtled toward the lowest part of the ravine. I lurched hard to the right to avoid following its path, my injured leg screaming in protest. Joe turned back quickly, concerned.

"Just clumsier than normal," I glowered, frustrated by the pain in my thigh and my own carelessness. Not wanting to waste the meager oxygen on speech, Joe resumed trudging.

We carried on through the stillness of the thinning atmosphere, no bird or animal noises to provide insight into the mood of the morning, no vegetation to cradle the ephemeral wind that kissed the slope, more a suggestion than a reality. Ragged breath was coaxed from each of us, human and equine alike. The simple task of inhaling, the work of alveoli deep within our lungs—exchanging oxygen for carbon dioxide—became an assertion of our right to live, to exist in an environment hostile to our intrusion. It is an ancient ritual, this effort of humans to know the unknowable, to breach the impenetrable. The primal trail was again a proving ground, but this time the immediate goal was science rather than survival.

Whether from the actual thinning of the air or the view of the world through oxygen-deprived eyes, all we saw seemed to enlarge and sharpen. The menacing presence of Salkantay slashed the infinite blue above. The harsh slopes below grew darker, the intensity of their blackness threatening to become abyssal. Even the aqua of the glacial lake turned brittle, a glassy shield that seemed like it might fracture upon impact instead of parting in ripples. No plants, trees, or even grasses grew here. Nothing lived. Implacable rock met insubstantial air met surreal water, a fortress of nature warning away life.

My feet halted, responding to something subconscious. I swiveled left, my gaze pulled as if by magnetic attraction toward a horizontal unzipping in the snowy façade of Savage Mountain. From left to right, a black zig-zagging rift sped across the vast whiteness. Before I could exclaim, the sound reached us. A great breaking, a massive cracking split the cliff face apart. We traded places with the ice, a group of travelers frozen little more than a stone's throw away while millions of pounds of solid water dove downward on a mission.

What began as a coherent mass divided and multiplied, first tens, then hundreds, then thousands of pieces of impatient ice, doing the math to reach terminal velocity on the way to union with its liquid cousin below. Sound waves raced behind, unable to keep up with the kamikaze strike of the avalanche. We stood silent as frost streamers formed glittery tails punctuating the fall of the debris. A stadium's worth of frozen rubble had come to a violent rest at Salkantay's feet, and still we waited. The valley reverberated with the dying call of a glacier, another victim of the changing climate we were working to understand. We paid our respects and turned again toward Inka Chiriasqua.

▲

We reached Sisaypampa late in the day. Setting up camp was our priority, since night steals over the landscape quickly in the Andes. Our wranglers and Gerardo set up the cook tent and secured the pack animals while Joe and I arranged our tent and equipment. As we finished, I caught sight of a skinny, tall tent popping up about 100 feet from the main camp area. Joe explained that it was our bathroom tent. Even in one of the most remote places I'd ever visited, our human desire for privacy when answering the call of nature was accommodated. Gerardo came over, apologizing that the bathroom tent's zipper was broken. They had the presence of mind to put the flap facing away from camp, so we'd be using the facilities with one hell of a view and hopefully in peace. The plan was for all of us to share a communal "cathole" until it was full, and then dig a fresh one. This was new to me, so I headed over to scope it out while Joe went to help the other men with dinner preparations.

The cathole was pristine, about 6 inches in diameter and dug about a foot down into the glacial till, which is a mix of soil and rocks deposited by glaciers hundreds of years before. A small shovel rested against one corner of the tent, and a pile of exhumed till sat close by the hole. Satisfied with the setup, I stepped back into the gathering dusk and headed for the light of the cook tent. The final vestiges of daylight glinted off Salkantay's icy summit thousands of feet above our camp. I swung

my arms and inhaled deeply, testing my body against the mountain air. It felt sharp, flooding my lungs with cold and a slim
hint of malice. The warning was clear—enjoy the splendor and
majesty but stay alert. This was not a place that would tolerate
human error.

▲

Our group settled into a routine over the next few days—as much
of a routine as we could claw out on the side of one of the tallest
peaks in the Andes. Our mornings began with coca leaf tea in
the cook tent. I'd seen people fall victim to mild altitude sickness
before, and it is a miserable experience for the afflicted person as
well as their teammates. Symptoms are alleviated by a descent
to lower elevations, which was not something our group could
have accomplished in a reasonable amount of time. I judged
it prudent to drink the coca tea and chew the occasional coca
leaf as preventative measures, remembering that the FBI has no
jurisdiction in Peru and what my parents didn't know was better
for the expedition.

After a consistently delicious hot breakfast courtesy of Gerardo, we loaded our packs for the day. Our gear included
cream-colored sturdy cloth sample bags, my trusty rock hammer,
Joe's sledgehammer, mallet, and chisels, a handheld Garmin GPS
unit, a hand lens, a Brunton geological compass, an inclinometer,
and the field notebook. We shoved sack lunches, extra layers of

clothing, sunscreen, lip balm, mini first-aid kits, 3 liters of water each, and our cameras into our daypacks. Having all the gear necessary for each day is key to a successful expedition. No one wants to hike for hours to get to a remote location only to find that they forgot the field notebook or GPS back at camp. Improvising in the field is essential since the unexpected becomes the norm when you work in wild places, but collecting rock samples without a rock hammer or marking sample locations without a GPS renders the work impossible.

We spent the first several days relatively close to camp collecting samples from numerous nearby sites in the valley. We were also trying to stress my injured leg as little as possible while still accomplishing the job. My flexibility wasn't returning, and the cold temperatures made it difficult to loosen the muscles.

The geologic targets of our activity were moraines—ridges of debris that form as a glacier pushes material during its expansion phase. Glaciers spread across the landscape almost inexorably, carving out U-shaped valleys. The sheer weight of ice destroys rock formations, carrying broken pieces of mountains to new resting places, sometimes hundreds of miles from the glacier's origin point. Moraines are physical evidence of the extent of a glacier, remaining behind like a cicada shell, proof that the inner area was inhabited by something no longer present.

Joe and I quickly fell into an effective working rhythm. Taking detailed notes is a best practice for geologic fieldwork. If you lose your camera, or the images become corrupted, having good field notes can mean salvaging hundreds of hours of work. Notes also preserve elements of the journey that would otherwise slip out of mind.

A good rock description can tell a reader not only about the rock itself, but about the environment in which the rock was deposited or formed. One of my professors at CSULA, Pedro Ramirez, set exacting standards for his students—receiving an A from him was a challenge. The experience stuck with me, and I have emphasized writing precise, detailed descriptions of rocks and sample locations throughout my career. A successful rock description can transport a reader anywhere in the world. I find immense satisfaction in scrawling pages of information in rain, snow, and under the glaring sun, filling the lines of yellow hard-backed Rite in the Rain notebooks, memorializing the intricacies of rocks from around the globe for scientific posterity. With the right description, a geologist who has never visited the area can understand enough about the rock to deduce the conditions that existed when that rock came to be in the location where it was found, even without any pictures. It's like creating a highly detailed description of a crime suspect. Photos help, but the notes serve as backup and vice versa.

Our process in Soraypampa quickly became well-oiled. Joe, being nimbler and the lead scientist, would scout for potential sample sites. Once he identified a spot, I lumbered over and pulled off my pack. Despite the long underwear beneath my cargo pants, I dreaded the bite of the cold granite boulders I used as seating. Legible field notes required a stable base, so I gritted my teeth and waited for my butt to equalize with the frigid rocks. Once seated, I wrote the unique sample identification number, the latitude, longitude, and elevation of the sample location, and the accuracy of the GPS unit for that location. GPS measurements all have some degree of error and knowing how far off a location might be could mean the difference between finding a site again and losing it forever. Our GPS unit was typically accurate to within 2 meters, or 6.56 feet. Some very expensive units can be accurate to within a centimeter, but for optimum performance we needed to be away from any overhead obstructions.

After we selected a site, Joe set to work breaking off about 2 pounds of rock to take back to the United States with us. He had a sledgehammer, mallet, and assorted chisels to get the job done, and heavy leather gloves and safety goggles were a necessity. This is no easy task, as granite is a very durable, resistant rock. Breaking enough rock free from a rounded boulder requires finesse and no small amount of brute force. While Joe struggled to wrangle samples into submission, I cracked into the rock descriptions.

## Field Journal

JULY 18, 2010

Boulder on crest of 5th ridge of the middle group of moraines. Surrounded with solid base that tapers uphill. Sample surface

retains sheen in sunlight that indicates vestiges of concordant
surfaces & original glacial smoothing. Mineral etching & joint-
ing evident. Spalling from sides likely. Boulder is ~110cm on
uphill side, ~150cm on downhill, sample is ~140cm. Sample is
~15% biotite, ~45% K-spar, ~20% plag, ~20% quartz. Large
crystals evident. Samples came from outer edge (~150cm) &
inner top (~140cm).

Not exactly Tolstoy, but detailed enough to give a geologist a
good idea of the sample's characteristics. The rock description
was the penultimate component of each entry, however. Since
we were collecting rocks to measure bombardment with cosmic
rays, we also needed to record whether any surrounding features
shielded the sample location from the rays. In our Peruvian location, we had to account for Nevado Salkantay itself, the other
mountains that soared skyward in all directions, and any boulders or moraines that protruded from the ground and could have
obstructed the sample from direct radiation. I used a handheld
compass and clinometer to locate the azimuth of each obstruction (the location on a 360° circle), and then determined the
altitude of the obstruction (the number of degrees it rose above
the horizon). Salkantay was our highest, topping out at 35° at
one sample location. This meant that we would need to calculate
the amount of radiation the sample had received and then sub-
tract the amount that never reached the rock due to the shielding

provided by Salkantay. Drawn out, my shielding records make spartan versions of compasses, interesting only within the narrow confines of our study.

This wasn't the glamorous part of science, but it was essential. Data were meaningless if even one element of the overall equation was neglected. Failing to properly record a sample's shielding would render the whole sample useless for our study. This type of fieldwork was one part brute force and one part attention to detail. On our first full day of work, we collected seven samples from the moraines sprawling out from Salkantay's base.

▲

The smell of cooking food welcomed us back to camp. Gerardo had prepared hot water for tea, and I decided to have a cup of coca leaf tea first, before my bedtime cup of chamomile, better known as *manzanilla* in Spanish. Warmth from the tin camp cup flooded my fingers, defrosting them palms-first. Andean night was swift and absolute. Our kitchen tent was the warmest spot around, a beacon nestled in the dark, silent valley of the Incas. Our wranglers pushed the tent flaps aside and entered, grabbing their soup before sitting on low stools at the far end of the tent, slurping wordlessly. Gerardo passed bowls to me and Joe, and we gratefully downed the contents. The hot liquids merged in my stomach to do battle with the twin challenges of cold and hard work–induced hunger. Bliss. Due to his work on the Inca Trail, Gerardo was fully prepared for people who had spent all day exerting themselves. While the others had something meaty, my meal included generous portions of soft, flavorful beans, fluffy rice, and vegetables with some amazing cheese. I maintain that the best food is (1) eaten after a long day in the field, and (2) cooked by someone else. Both of those criteria were satisfied, and I declared this was dinner fit for a king.

As we finished our meal, Joe and I were both giving side-eye to our younger wrangler's clothes. He was in his early twenties and had straight black hair poking out from under a beanie with llamas and traditional Peruano geometric knitted designs. His eyes were creased from the sun, which is brutal at the altitudes where he lived and worked. He wore a jean jacket over a knitted sweater featuring more Peruvian designs. His jeans were black and embedded with dirt from previous trips. The shock came from his footwear. His bare toes were visible in his black rubber sandals. I was already wondering how I would stay warm that night, and I had a down-filled sleeping bag, long underwear, and both silk and wool socks. I resolved to discuss the matter privately with Joe. I knew that the Quechua are well-accustomed to the harshness of the Andes, but not wearing socks in below-freezing temperatures just seemed painful.

As soon as the wranglers finished, they headed back into the night to tend to the animals. Their tents, and Gerardo's, were set a few hundred feet from ours. I couldn't talk to them directly due to the language barrier, and that made me sad. Not only did I want to hear about their lives and thoughts about our work, but I also wanted a diversity of conversation partners since this was a lengthy trip. The same corny jokes only work so many times before travel companions contemplate the odds of anyone ever finding the body. Joe and I had a good rapport, however, so my focus remained on not allowing my injury to impede our research. So far, so good, but we were just getting started.

▲

The morning of the snowstorm was an exception to our daily routine. I awakened during the night with the vague urge to pee, but after opening the tent flap and peering out into a zero-visibility snowfall, I concluded that holding it until morning was the best choice. Making decisions while still more unconscious than not is a bad policy, so when I woke a few hours later to muted daylight, I rushed through adding outer garments to my sleep attire, racing against my unhappy bladder. I stumbled outside, narrowly avoiding a full faceplant into at least 6 inches of wet, heavy snow that had cascaded from the sides of our tent onto the ground in front of my entrance flap. Glasses askew on my nose, I cursed as wet snow invaded the dry sanctuary of my boots. I wasn't being careful, and now I would have to give this set of socks to Gerardo for drying over the fire.

# Field Journal

*Joe—It snowed all night and we woke up to find several inches of slush covering the tent and everything else—snow/sleet/rain continued throughout the morning, so today has been declared a "snow day."*

Continued in my hand:

*Jess—We have been in the tent all day.* AAAAAH!

Relieved, I exited the bathroom tent and tried to compose myself after the unexpectedly snowy, urgent awakening. I turned right to gaze down the valley, away from Salkantay. The sharp air filled my nostrils, more a feeling than a smell. I felt water vapor crystallize inside my nose, individual hairs hardening into tiny icicles on the inhale, then thawing on the exhale. A silken white blanket enveloped the valley floor and continued, unbroken, up the slopes on either side. The vivid blue sky in contrast with the luxurious whiteness below was like nothing I had ever seen before. Pure, utter, and absolute sky unsullied by contrails, clouds, or birds. It seemed to pool, its depths—or perhaps heights—inscrutable and endless. The annoyance of the rush to the bathroom slipped away, and I returned to our tent, determined to capture the dazzling mid-July snow in a photo. I jammed my contacts into my eyes, afraid clouds would appear from behind Salkantay and mar the perfection.

The weather was on my side, and I shot one of my all-time favorite expedition photos that morning. I had finished shooting when Joe returned from the cook tent, trudging through snow that was on its way to becoming slush. We talked, deciding that the snow would make work much more difficult. I enjoyed my first snow day since leaving New England after college five years earlier. We used the day for writing, talking, and resting. Fortunately, Joe had built time into the schedule for us to rest or deal with problems. Good expedition planning came through yet again.

The sun's rays were working hard to turn the night's snowfall into afternoon puddles of slush. Joe and I discussed our wrangler's lack of adequate footwear when we glimpsed him tending to the pack animals, who were pawing through piles of slush to reach grass, their food source the only thing aside from lichen and moss able to grow in our valley. We had several pairs of thick wool socks in our packs, and we decided we should offer him a pair. My injury was one strike against us, and both compassion and practicality spurred us to act. Joe brought a pair of his socks when we went over to the cook tent for dinner. We were a little hesitant since we didn't want to offend, but when Gerardo translated the offer he grinned, showing us the first big smile we'd seen from him. He pocketed the socks while the older wrangler eyed him, impassive.

After dinner we somehow broached the subject of sports. Gerardo was trying to include the wranglers in our discussion, flipping between Quechua and Spanish. Joe is not a big sports fan, so it fell to me to explain American football to three Peruanos, two of whom had zero concept of the game. I ended up explaining that major cities had teams, and even going into minutiae like the NFC versus AFC, field goals, and sacks. Drawing comparisons with soccer seemed to be my most successful tactic, so I did that often. The highlight for me was describing my dad's favorite team as *los Gigantes de Nueva York*. I didn't even try to explain the Redskins.

As I walked back to our tent to get ready for bed, the snowy flanks of Salkantay formed a glistening stage for the almost-full moon. The mules and horse grazed in silver moonlight, the steam of their breath dissipating as soon as it left their nostrils. Sisaypampa lay still, painted in silver, cerulean, and cobalt. Salkantay loomed in the background, unmoved by the quiet stirrings at its feet.

▲

The next morning we returned to our workday routine of collecting samples and eating a lengthy dinner near the fire, drawing out our time in the warmth and community of the cook tent. Since we had worked, our conversations were not as involved as those on the snow day. I didn't want to exhaust Gerardo's goodwill by requiring more translations, but Joe and I both happily noticed that the younger wrangler's feet were now snug in Joe's socks.

Before turning in, Joe and I each took a few swigs from a bottle of imitation Baileys that we had purchased back in Cusco to share during the long, cold expedition nights. The cheap liquor burned pleasantly as I drifted off to sleep, but Joe elbowed me awake after what felt like only a few minutes. His whisper was ragged and urgent as he told me that one of the wranglers had roused him because armed horse thieves were in the camp. I blinked, fumbling for my glasses. Out of habit, I had brought my rock hammer into the tent with me. Remembering that I had my favorite 4-inch-long pocket knife in my bag, I handed Joe the hammer. Our wrangler's instructions had been to be silent and stay in the tent, so we complied. My glasses fogged repeatedly, something that was generally a minor annoyance becoming a potential impediment to my self-defense—if it came to that. I flicked the knife open, gritted my teeth, and made a mental note to get Lasik surgery as soon as I got a paying job.

Our tent was on the opposite end of Sisaypampa from where the livestock had been grazing, and our tent's rainfly meant we had zero visibility. My breathing was shallow, my heart rate higher than normal but not racing. I couldn't hear anything that indicated it was time to panic. The cold air combined with the leftover slush from the day before to ensure any approaching footsteps would crunch noticeably on the grass. Joe's shallow breaths were the only sound that reached me.

Mentally, I began to review potential scenarios and what actions I would take. Until this point in the expedition, I hadn't thought about being the only female. The only notable difference between me and other expedition members was my injured leg. I hoped that the horse thieves didn't know I was there but began steeling myself to stab the living shit out of anyone who tried to breach our tent. A 4-inch knife can easily become lethal. This knife was an old friend, and I had carried it regularly since my college years. The handle fit my hand, the balance and heft ideal for what I hoped I wouldn't have to do. I reasoned that I had the half-full bottle of fake Baileys to use as an initial weapon and could follow it with several quick slashes and a final sharp jab. I had been assaulted once before and was unable to fight back that time. The remembrance split my thoughts into two distinct tracks, and I assessed the odds were in favor of this all ending very, very badly. I was hoping for the best and planning for the worst, which is up there with the best advice I'd ever received. Fear was not an option, only a tiny bit of optimism and a giant knot of resolve twisting around in my gut.

After an impossibly long stretch in that state of suspended animation, a metallic *click* reached us from across the camp. Electrified, we snuck wide-eyed glances at each other, the most movement we'd risked since I had passed the hammer to Joe. I held my breath, straining to catch any other sound at all. Nothing. Nothing. Nothing.

*Crunch-crunch-crunch-crunch-crunch*-CRUNCH-CRUNCH-CRUNCH.

"Toda está bien, amigos."

Gerardo.

"Oh my god," we both exhaled.

We bolted up, unzipping the tent and spilling out into the frigid night, trying to get details from Gerardo in Spanish while our adrenal systems purged the hormones as quickly as they'd released them. I had never been so eager to leave a warm tent.

Over hot tea in the cook tent, Gerardo explained the thieves had used the full moon to scope out our pack animals the night before, and they had returned with—at the very least—thieving intent that night. He told us there had been some sort of a struggle, which had sounded indistinct to us as we waited, coiled and taut, inside our tent. It may have been either an intentional obfuscation on Gerardo's part, or perhaps just a slight language difficulty, but the conversation didn't leave us entirely clear on what had transpired.

Joe and I were unaware the older wrangler had a rifle, and him chambering the shells produced the metallic noise we heard. Apparently, the sight of the gun, plus him, the younger wrangler, and Gerardo all awake and alert proved too much of a challenge for the thieves, who disappeared into the end of the valley where we had yet to explore. Other details were hard to come by, but our relief was overwhelming. When we tried to get back to sleep, my mind went to places I hadn't allowed it to during the actual incident. Torn tent shreds flapping in the night, crimson pools on sleeping bags, Joe with a bullet hole in his skull . . . and then my brain refused to let me follow that path any further. Visualizing my own worst-case scenario would have pushed me into a hole I couldn't have climbed out of while still in the mountains.

If our wrangler had not had the foresight to bring the rifle, I don't know how the night would have turned out. I slept restlessly, and when morning came I was relieved to see the light of the new day. I eyed Salkantay, wondering if the mountain had any other savage situations in store.

The next morning, Gerardo made clear we should leave our quiet valley, since he worried the thieves might not take well to being deterred and return that night with reinforcements in the form of arms or manpower. Rather than risk outright confrontation with a group of frustrated thieves, Joe decided that the data we'd already collected would suffice. We broke camp, the slight disappointment of leaving sooner than planned soothed by the impassive bulk of Salkantay looming silently behind us while we worked, seeming to grant a reprieve from the full force of its savagery. We would make it back to Cusco with no permanent damage and a new appreciation for the value of a solid field team.

# Sierra Madre Occidental

## THE MOTHER MOUNTAINS

DECEMBER 2010 THROUGH MARCH 2011, MEXICO

I SHIFTED THE hand holding my stiff yellow-backed notebook, trying to find an angle that muted the hammering deep within my left shoulder. Momentarily successful, I jotted down more information about the crumbly outcrop of weathered andesite rock poking out from underneath the bramble of rural Sinaloa, Mexico. Hunting ancient volcanoes has none of the glamor of Harrison Ford diving into ruins of civilizations long gone, punching Nazis, and stealing idols in the name of history and science. However, volcano hunters *do* encounter Indiana Jones levels of danger, and not always from our subject matter.

On this expedition I was nursing a surgically repaired shoulder that required months of wearing an unwieldy Velcro brace. An old dislocation had recently recurred and the only repair option had been a rushed surgery, which seemed to have gone well despite the tiny amount of time I had to arrange it. My left arm was immobilized down to the wrist by the straps pinning it to my side and stomach. This situation would have rendered my work impossible were it not for Carlos. Somehow, I had managed to convince my brand-new husband that our honeymoon should involve sledgehammering hundreds of rocks throughout one of the most dangerous states in Mexico . . . for science! To help my case, I found a tiny fishing village with casitas for rent, just steps from the ocean. It seemed like the perfect base for three months of piecing together the geologic history of when the Baja Peninsula split off from the rest of Mexico, creating the Gulf of California.

I bent forward for a closer look at the andesite, a volcanic rock associated with violent eruptions due to its high silica content—and, consequently, those eruptions are connected with high-viscosity, sticky, explosive lavas. This rock was millions of years old, and I was hunting zircons—tiny minerals that could help me attach more precise dates to the eruption that produced the rock using the trace uranium and thorium they carried. Dating eruptions is important to geologic research, because understanding the past gives us reference points for what can happen in the future. At that moment, though, I didn't care about the historic significance of the rock. Bending forward wrenched something in the fragile interior of my shoulder, and the familiar thumping returned.

Before I verbalized my pain to Carlos, who was busy bludgeoning some potential samples into portable fragments, the low noise of an engine reached me. I told him to stop and listen. He did, squinting at nothing. He nodded once, then grabbed the sample he was working on and the sledgehammer and headed for our bright red rental Jeep. I spun and made my way to the passenger side, each hurried step jarring my shoulder.

We had barely shut the Jeep's doors when a white Dodge Durango with spinning rims and blackout window tint everywhere except the windshield materialized around a bend in the road. We were 15 miles from the nearest paved road, and the popped collars and aviator sunglasses sported by the Durango's driver and passenger telegraphed their affiliation with a Sinaloa cartel. We looked down, trying to appear as nonthreatening as possible. I had planned for this by having magnets made with the official crest of la Universidad Nacional Autónoma de México, or UNAM, Mexico's prestigious public university and home institution of one of my research collaborators. The narcos rolled up next to our parked Jeep, and I looked away. Carlos, with his Mexican and Colombian ancestry, usually raised little interest with the locals. My red hair and blatant American appearance always garnered looks and curiosity in Mexico, so I hoped my bandana was enough to obscure that I obviously wasn't from around there.

After they passed, Carlos said they'd checked out the magnets on the Jeep as they rolled by. I asked if he could see if there were people in the back of the suv, but he couldn't tell due to the window tint. As soon as they disappeared down the road, we drove away, deeper into the mountains ahead for half a mile or so. I wanted to put distance between us and them. The branches on either side of the road stopped abruptly and the road widened into the entrance to a makeshift village. Wooden shacks leaned precariously behind barbed-wire strands dropped casually on the ground. A dog barked and was soon joined by others trying to raise the alarm. This was very likely a marijuana farming operation, and we didn't want to stay around long enough to confirm that hypothesis. Carlos pulled the Jeep into a sharp turn and hit the gas, and we stopped once the clearing was gone from the rearview mirror.

We inhaled deeply, and I noticed a nice outcrop of andesite just outside of my window, along a steep bank that vanished upward into a tangled mess of naked winter branches. I prepared to get out and look, and paused.

"Hey, where's the rock hammer?" I frowned, straining to see behind me since turning to my left was impossible with that damned shoulder brace.

"I thought you had it," Carlos responded.

"No, I only had the notebook and my pencil," I said with growing unease.

He shifted into drive and eased off the Jeep's clutch as we started rolling forward. To a field geologist, a good rock hammer is indispensable. Seeing the inside of rocks, the parts free from the ravages of weather and time, is how we discover their true nature. Since we don't have x-ray vision, the rock hammer makes understanding the heart of those often very solid objects possible. I often joked that if I couldn't fix a problem with my rock hammer, it couldn't be fixed. I'd used mine for sampling molten lava, prying apart pieces of mountain lion scat, hanging pictures at home, throwing at hay bales for entertainment à la lumberjack axe throwing, and as a self-defense tool. It was the most versatile piece of scientific equipment I had, and it was critical to the two months of work still ahead of us in the remote mountains of Mexico's Sierra Madre Occidental.

What set this hammer apart from generic hardware store models was its construction. Mine had a pick end and a more traditional hammer end, with the head alone weighing 20 ounces. The handle and the head were made of a single piece of drop-forged steel, which means there's no chance of the head separating from the handle unless something has gone dramatically, terribly wrong. If your rock hammer was dead, so were you. That made it perfect for banging on the hardest rocks imaginable in relative safety. I had chosen a hammer with a black-and-yellow rubber grip over the almost ubiquitous blue-handled Estwing hammers because I liked the easy visibility of the cheery yellow against rocky backgrounds.

As we approached our previous sampling site, I scanned the mottled gray base of the outcrop for that bumblebee-colored handle. Nothing. Worried, I asked Carlos if he saw it. Negative. Frenzied thoughts erupted in my mind. Our base of operations was over an hour and a half from Mazatlán, the closest major city, where I was certain no stores were selling rock hammers. Ordering one from the United States would take weeks, and I had no idea if a package would even make it to me in the tiny fishing village. The sledgehammer could do a reasonable job of creating big chunks of rock, but to get them down to the size I needed for samples would be impossible without the dexterity of the rock hammer. I wasn't going to ship scores of 5-plus-pound rocks to the lab.

"They took it. Those assholes took it."

I was calm, but Carlos heard the steel under my words. We had pulled around a curve in the road before they'd passed our sample site, so I reasoned the narcos must have stopped after our Jeep was out of view. He looked at me, then back at the rugged dirt road leading down out of the mountains. He stepped on the gas and the Jeep's engine growled, already in pursuit. The road twisted and dropped irregularly, rendering it impossible for all four Jeep tires to stay in contact with solid ground at the same time. We shot ahead, bouncing into the wild forest, vegetation cut through with the rich gold of the late afternoon sun. For my first car chase, it was a beautiful one.

▲

My bad shoulder screamed with every jounce of the stiff suspension, and I braced my good arm on the Jeep's hardtop roof, working to minimize the agony. Carlos radiated focused adrenaline, hands gripping the steering wheel, his dark eyes scanning the rough terrain ahead. Not for the first time on this expedition, I was grateful my project was in a Spanish-speaking country so I could more easily justify having my new husband as my research assistant. There was no one else I would rather be with while racing down the remains of an ancient volcano in pursuit of dangerous gang members. I turned my gaze back to the road ahead, watching as tree branches whipped the windshield and sunlight flashed through the trees with the effect of a frantic strobe light. I wondered if the Jeep's suspension would hold and realized we were about to find out.

It took about ten minutes of hard driving to reach the flat terrain in front of the small village we'd passed on our way up. As we emerged from the tree cover on the mountainside, I could see a faint dust cloud I assumed was from the Durango we were chasing. As we drew closer, an old beater of a truck appeared on the road ahead of us, rolling along at a leisurely 15 miles per hour. There had been no turn-offs, so I told Carlos our target had probably overtaken the slowpokes since the road was now wide enough to accommodate two vehicles side by side. He nodded and accelerated past the truck, and I stuck my hand out the window and waved an apology for the rude passing. Carlos spotted the Durango ahead before I did, and it was traveling faster than the beater truck but slower than us.

As we pulled even with them, it occurred to me they could have a small arsenal of weapons ready that we'd never see, hidden behind that black glass. I gulped and pasted on what I hoped was my friendliest, least threatening grin as we drew near. I made the old-fashioned circular motion for "roll down your window," and hoped the driver was considering my request and not chambering a bullet. The Durango slowed to a stop, and we matched their deceleration. I kept grinning as the words started to tumble out once the driver's window began lowering.

"Hola señores, soy geóloga. Necessito mi martillo. Es negro y amarillo, y es muy importante para mi trabajo." Despite my polite words, my actual thoughts were more forceful.

*Give me my damn hammer back and don't shoot me, you assholes. You don't even know what to do with it!*

I could see both the driver's and passenger's expressions change behind their aviator sunglasses. Their eyebrows rose upward, and it looked like they were torn between responding or laughing. I couldn't see their hands, but my entire body was strung as tight as a bow ready to fire. Either they would be nice, reasonable drug runners and give me my hammer, or Carlos and I would be the first geology research team stupid enough to get themselves disappeared by an international drug cartel over a $30 rock hammer.

"¿Sí, es verdad?" The driver's face and tone of voice were conveying disbelief more than anything else—not disbelief about my profession, but about the situation. Here was an UNAM Jeep with a Spanish-speaking redheaded American and a silent Latino, and these *pendejos* had the audacity to chase them down a little-used dirt road far from all possible help just to ask for a hammer back. This was not something they'd planned for. Since he'd just asked if what I said was true, I decided to elaborate.

"Sí, estoy estudiando los volcanes cerca de aquí," I went on, deciding it was unlikely any of them knew the mountains where they did business were volcanic.

Bingo. The driver's eyebrows rose so high they looked like they were making a break for his hairline, and his response made it hard for me to keep my idiotic grin in place.

"¿Hay peligro?" He sounded alarmed, and I saw the passenger lean forward, riveted to learn if they were all in danger of fiery demise.

I felt some of the tension in my body dissipate, as I realized the situation was now as close to under control as it could be. I used my best reassuring voice.

"No, no, todos volcanes aquí son muy viejos. No pueden explotar. Estoy estudiando los minerales dentro de las piedras para evitar una erupción en el futuro." By easing their fears and explaining what I was doing (and only exaggerating the importance of the work enough for it to seem critical), I hoped they'd realize the UNAM magnet was all the proof they needed that we were just harmless rock nerds. I didn't expect the driver's response.

"¿Pero hay oro aquí?" *Of course* he wanted to know if there was gold. He was in a drug cartel.

"¡No, no, yo deseo!" I laughed, hopeful that acknowledging humanity's shared greed was the right decision. The flicker of a smile crossed the driver's face, and he turned away from me, toward the back seat. I could hear him saying something, but I couldn't make out the words. A moment later, he raised his hand and I could see the familiar yellow-and-black handle in his grip, the head pointing down.

"¿Es este tu martillo?" He asked the question innocently, as though a large rock hammer had just magically appeared in his vehicle with no explanation. I was not about to criticize his delivery, so I nodded vigorously and began thanking him for finding it, explaining my work could now go on. Carlos noticed I was probably going to overdo it and leaned forward and thanked the driver himself. I took the rock hammer from the driver's outstretched arm, waved again, and rolled up the window. As soon as it closed Carlos pulled away from the Durango, slow enough to avoid choking them with our dust, but quickly enough to show we were out of their territory. I gripped the hammer tightly, giddy with relief and our continued good fortune. I never thought I would be so happy to get back onto a main highway, but the honest narcos had just proved there were exceptions to every rule.

▲

After a few weeks in Sinaloa we'd settled into a rough routine. Each day we drove to an area where I'd identified possible locations of outcrops of volcanic rock using geologic maps and satellite images. We'd take the Jeep on roads that were no more than faint traces cutting across rolling fields of windswept wild grasses, around small mountains I knew were the ancient cores of once-active volcanoes, and through towns where every cement and cinderblock house had nothing more than dirt for a floor. We passed old Sinaloan grandmothers, faces wizened bronze by years in the relentless sun, sweeping the dirt patch in front of their turquoise, lemon, or cotton-candy-pink block houses.

Tiny family cemetery plots jutted out from small hills or nestled in little depressions near the side of most roads, each having at least one statue of a *virgen*, her robes framing her out-stretched hands while her unseeing eyes watched us pass. The pre-Columbian polytheistic religion shoehorned into awkward matrimony with more formal Catholic iconography gave the local holy places an air of incompleteness, a seeming inability of the souls within to fully rest. Mexican Catholicism's bright colors and incorporation of elements of so-called pagan traditions stood out as very different from the staid Catholic traditions of my early years. Our carnival red Jeep made a stark contrast to the rusting trucks and tired sedans driven by the locals—this one with a bumper held on with baling twine, that one's mismatched body panels proving its Frankenstein origin story. And, oh, the unending games of *fútbol* in vacant lots, most kids with only a pair of shorts and a ragged T-shirt of their own, choreographing their love for the game amid the debris and the stray dogs and the endless dust.

Aside from two brief trips to Puerto Peñasco, a tourist haven just 60 miles from the U.S. border, I hadn't spent any time in Mexico. Thanks to a mother who spoke Spanish well enough to become one of the first female FBI agents under their language program, I wasn't worried about language barriers. My main challenge deep in the state of Sinaloa, miles from any tourist destinations, was not to stick out like the sorest of thumbs. Mazatlán, the closest city with an airport to my research area, was experiencing a slight decline thanks to the rise in cartel violence. We were also there in winter, so the vibrant city life that the cruise ship port town is known for was absent.

Most of our three months of work were spent closer to a town called Dimas, home to just 3,500 residents, which felt enormous and crowded after long days chasing ancient volcanoes in barely developed farmland or virtually uninhabited low mountains. Understanding how and when the volcanoes of the Sierra Madre Occidental erupted, and whether their volcanism was tied to the opening of the Gulf of California was the heart of my investigation. I had accepted a PhD position with a university in Brisbane, Queensland, Australia, and after arriving down under, turned around three weeks later to start my research on the volcanoes of Mexico. Carlos left his consulting job in Los Angeles to make the move with me, and we married just a few short weeks before we left for our new life. His work in finance and auditing was portable worldwide, so we were up for an adventure to kick off our new chapter. I found ways to ship my dog, cats, and Chevy Blazer, but my horse, Vash, was going to have to bide his time in a pasture in Texas until we had enough money to bring him over.

I had enough flexibility as a PhD-level researcher to choose my own approach to my fieldwork. After examining some maps and putting my Google skills to work, I'd found Barras de Piaxtla, a village with a name that even my native-Spanish-speaker husband struggled to pronounce. From what I could find online, it seemed to be known for a summer surf school and a small guesthouse compound called La Rosa de Las Barras. The proprietor was an American expatriate named Gail and the prices quoted for the little houses, or *casitas*, were within the budget of my research funds.

When Carlos and I arrived at La Rosa, we were astonished by the striking beauty of Gail's beachfront compound. The casitas themselves had thatched roofs, and the décor inside was simple and fit the location, but everything had a dramatic flair thanks to Gail's artistic efforts. Our casita had exterior walls made of machete-hewn wood while the interior walls were covered in bright white plaster. The kitchenette counter was made of natural rock with caramel-colored wood cabinetry. An outdoor shower next to the main dwelling area was walled off with more local wood for privacy. It was tiny, adorable, and the most unlikely working honeymoon spot we could have found. After long days

of scouring the countryside and breaking thousands of pounds of
rocks to collect samples it was a relief to come home to such luxury—especially considering that my usual field accommodations
were a tent and a sleeping bag on the ground.

Each day saw more of the fields, the brief inland villages, the
discovery of large boulders, Carlos shattering them into manage-
able pieces with the sledgehammer, and me cracking them down
into fist-sized chunks with the rock hammer. When I had what I
determined were samples that accurately represented the volcanic history of a given area, we crossed it off our list and moved on.
For lunch, we'd eat fresh, handmade tortillas purchased from one
of the many women selling them by the side of the larger roads
along with some beans I'd heat in their can on the dashboard of
the Jeep, the sun doing the cooking for me. In the evening, we'd
roll into La Rosa and I'd walk to the store while Carlos organized
samples. It was my nightly chore to try to coax more than one egg
from the shopkeeper, who would only sell one at a time so other
people in the village could also have one if needed.

On special occasions—when I was tired of cooking—we'd
walk to the one restaurant, which was the newly opened "Jardín
de Danny," named after the village's other American expat, who
ran the surf school in the summer and was trying to broaden his
share of the local economy in the off-season. All the tables were
outdoors, which was a practical choice for the mild weather and
a frugal one when it came to reducing construction expenses.
The menu was ambitious, featuring every Mexican dish I'd ever

heard of, plus others that were new to both me and Carlos. Typically, we'd order something from the very serious waitress who would then disappear inside to the kitchen. Several long minutes after our order, the waitress would reappear and inform us that our selections were unavailable. We repeated the process a few times until we reasoned it would be quicker to ask her what they *did* have instead of making her find out what they didn't. Unless Danny himself was drinking Pacíficos on the patio with some friends, we would be the only customers. It was peaceful, yet also akin to being among the only survivors of an apocalypse in paradise.

▲

By early February, we were in our third month and at ease in the little village. Gail had moved us into the only two-story casita on the property. Her house was up on the hill overlooking the whole village, and we only saw her once or twice a week. When she offered us the larger casita, we gladly accepted. It had a full kitchen, a small living room with a television that got a few channels via satellite dish, and a private balcony adjoining the upstairs bedroom with an unobstructed ocean view. We were the only guests in the entire compound for all but a few days of our stay, and Gail seemed to enjoy hearing about our work. She charged us the same rate for the two-story casita, so we felt lucky to get the upgrade.

We spent mornings exercising on the beach, and as my shoulder continued to heal from the surgery I began to jog slowly along the sandy strip between the water and La Rosa's property. Kanye West's "Stronger" was my encouragement to persevere through the deep, ever-present ache inside my shoulder. Three long months of being sedentary were strong motivators for me to seize whatever motion I was able, no matter how painful, and even the scars from the laparoscopy had transformed from angry red gashes to the shiny pink optimism of healing skin.

We had the occasional run-in with local narcos, like the time we were driving between fields full of baby vegetables, absorbed in trying to get as close to a potential sample location at the foot of a small mountain as we could with the car. A white pickup truck with dark window tint and a crew cab slowly rolled up parallel with us, but fortunately they were on Carlos's side of the Jeep. When the occupants began talking with him, they visibly relaxed as he pointed at the UNAM logo on the side of the Jeep, explaining how we were just scientists looking to study old volcanoes. Our experience with the hammer-stealing narcos had demonstrated the value of playing up the danger of the region's volcanics, and we employed that tactic whenever we interacted with the locals. While I didn't want to create false fear of volcanoes, I did want word to get around that we were scientists doing important work to protect the community. Any time a situation was potentially dangerous, I preferred to put my finger on the scale to tip the risk calculation in my favor.

One evening, after another long day of sample collecting, we swung through Dimas to buy gas and pick up a few groceries that weren't available at the tiny *mercado* in Barras de Piaxtla. As we were leaving town, we saw a black armored truck bearing the insignia of the federal police heading toward the town center. In the bed of the truck were several policemen clad in all black military-style gear, each with a black mask over his face and a semi-automatic weapon grasped tight. All of them were standing up, swaying with the motion of the vehicle as it navigated the road's curve. In the center of the group was a large gun mounted to the truck itself, with one policeman behind it, clutching its controls. I had no idea what type of weapon it was, but its power and lethality were apparent. I commented to Carlos that it must be a busy evening in Dimas.

Less than a minute after we passed the team of heavily armed police, we saw a large white pickup truck heading the same direction the cops were. The truck had some dents and scratches, but it was the men half-hanging out of its windows and clinging to its bed that were eye-catching. Every single one was holding a semi-automatic rifle, and this truck also had a massive weapon mounted in its bed. In a sharp contrast to the police crew that had preceded them, these men looked like they were agitated, poorly nourished, hunting for confrontation, and likely users of the drugs their bosses ran to the north. None of their outfits matched, and the only feature they all shared aside from the guns were bandanas pulled up over their noses and mouths, obscuring everything but their eyes. Our windows were down, and we heard them yelling and pounding the sides of the truck as we passed. We exchanged a worried look and got the hell out of Dimas, barely slowing down to take the curve onto the Culiacán-Mazatlán highway. We'd learned that when the narcos and police were in the same neighborhood, the only safe option was to leave.

We arrived home after dark and reheated leftovers before a quick trip to the handmade concrete hot tub Gail had constructed in the compound's courtyard. She had inlaid part of it with custom tile mosaic work, and it was nestled away from the view of the casitas. With no other guests in the compound, we were able to relax and enjoy the expansive tub. The tension of a long day of driving and breaking rocks washed away, we headed back to our casita and upstairs to bed.

Shortly after midnight, I was awakened by loud voices and shrieks of high-pitched laughter. Disoriented, since the only noise I ever heard in the middle of the night was the soft murmur of waves on the beach, I reached for my glasses and cell phone. I confirmed it was indeed the middle of the night, and then I nudged Carlos awake.

"Hey, wake up. There are people here," I whispered urgently.

"Huh? What?" Carlos was a sound sleeper, and we often joked he could sleep through a bombing. It took another banshee-like laugh from the outside to get his full attention. He sat up in bed, and we saw the beam of a strong flashlight sweeping across the windows of the casita. I appreciated our perch on the second floor, since judging by the number of voices there were at least a dozen people outside. Carlos got up to investigate, and we left the lights off. The only thing resembling a weapon was my rock hammer, so I handed it to him. He asked what I would do if something happened to him, and I told him I'd use a curtain rod if it came down to it. As he ventured downstairs, rock hammer in hand, I fervently hoped it would not actually come down to me using my one good arm and a flimsy curtain rod to fend off our nocturnal visitors.

I crept over to the windows nearest the voices and could make out Carlos's familiar tone engaged with a deeper, harsher-sounding one. I pressed my cheek hard into the glass, straining to get a view of who he was conversing with. Above the high wooden walls of the compound, I could make out the shape of a man with a very large rifle. Judging by his height relative to the compound walls, which were at least 7 or 8 feet tall, he must have been standing on something. My mind snapped back to the men in the bed of the trucks we'd seen in Dimas, and I realized these people were definitely not cops. That left the cartel as their only other likely employment, because there weren't too many lines of work that required armed guards after midnight in the most remote fishing village in Sinaloa. As my eyes swept the area visible from the window, I could see more men with large weapons patrolling the compound aggressively. I wasn't sure how they had gained access, since we locked the side gate after we had arrived, and Gail kept the main gate locked during the off-season.

A few long minutes ticked by. Then I heard Carlos's voice moving away from me, toward the back of the compound where the shrieking laughs were originating. I scooted across the dark room to the opposite side of the casita, where a window showed me only a shadow version of what I quickly recognized were people in the huge hot tub. The floodlight near the tub projected the images on the far wall of the compound, but it was confusing. With effort, I counted four separate shadows I guessed were in the tub. I could no longer hear Carlos. I noticed my heart skipping beats, which happened when I was either excited or stressed. After a moment of watching the shadows, I squeezed my eyes shut and tried to listen to the men walking around the casita. I could hear bursts of static followed by a few terse words, but I couldn't discern what was said. The static was enough to tell me they were communicating by radio, which confirmed my cartel suspicions. Every minute that passed without Carlos returning ratcheted up the pressure in my stomach, anxiety threatening a physical manifestation.

I heard footsteps crunch over the gravel, and then the front door of the casita opened and closed. Frozen in place, I waited until a faint beam of light illuminated the stairwell. It was too weak to be a flashlight and seemed more like the LED light from a cell phone. As soon as the light turned the corner, I recognized Carlos's outline and exhaled sharply. I hadn't noticed I had stopped breathing when the door had opened. He moved up the rest of the stairs as soundlessly as he could and sat on the bed. I joined him, and he put his arm around me. We kissed, and in a harsher whisper than I'd intended I demanded to know what happened.

As he related the story, I felt my eyes widen. I was incredulous, but overall it fit with our experience in Sinaloa thus far. When he went downstairs, cell phone flashlight on, he'd practically opened the door into one of the armed cartel members. The lookout had asked him who he was and what he was doing there, and Carlos explained he was staying in the two-story casita. The gang member then asked to see if he had any weapons, and Carlos showed him the hammer. Judging it no match for his rifle, the lookout radioed something to an unseen teammate and then told Carlos they were going to see *el jefe*—the boss. He led Carlos to the hot tub, where a man in his early thirties with a body soft from lack of physical activity frolicked with three surgically enhanced tittering women. All of them were nude, and empty beer bottles lined the hot tub rim. At the arrival of Carlos and the lookout, the boss man looked annoyed. The women started talking to each other while the drunk gang boss interrogated Carlos.

After hearing Carlos's explanation that the two-story casita was occupied, the boss grew angry and insisted it was *his* house. He said he always stayed there, and no one else could use it except for him. Carlos did his best to placate the man, offering to call Gail and see what could be arranged. Several attempts later, the boss agreed to talk with Gail. Carlos only heard the drunk cartel leader's side of the phone conversation, but he said

it sounded like a kid being told he couldn't play with his favorite
toy. When he hung up, he scowled at Carlos before turning back
to his lady friends, swigging a beer and grasping for the closest
set of breasts. The lookout nudged Carlos with the butt of the
rifle, and they returned to the casita.

I had no idea what Gail said to reason with the boss, but Carlos was confident we'd be fine for the night. We returned to bed,
but my sleep was more like a struggle than actual rest. When
our alarm went off at 7 a.m., the compound was deserted. Even
the beer bottles were gone, and if Carlos hadn't corroborated the
experience, I would happily have written it off as a bad dream.
As we pulled the Jeep out of the compound, a convoy of military
vehicles approached. They dominated the tiny central thorough-
fare of the village, ten trucks pulling up outside the compound's
wooden walls. The man in the passenger seat of the first vehicle motioned for us to stop, and Carlos obliged. He approached
the driver's window of the Jeep, full-page color photograph of a
person in his hands. With zero formality, he thrust the picture
at Carlos and asked if he'd seen the man, who seemed to be a
high-priority cartel boss. At that moment, I understood my new
husband could be a gifted liar when necessary.

Just a few more questions and the man was convinced we
had no idea that drug cartels even existed, let alone were active
in Sinaloa. He waved us off, and the trucks began to disgorge
their loads of armed men. I watched in the side view mirror as
they began to stalk around the walls of the compound, and saw

Gail's old pickup arrive from her house up on the hill. If this sweet, artistic American expat could handle the *jefe* of a notorious drug cartel, I was certain she'd eat Mexican law enforcement for breakfast. Suddenly, another long day off-roading in search of rocks seemed like just the right amount of excitement to me.

▲

When the fieldwork wrapped up a few weeks later, the news reports indicated the situation was worsening in Sinaloa. We spent a few nights in Mazatlán, where we heard gunshots outside of our hotel room—literally outside, as in next to our window— and we spent a solid hour shivering in the unheated bathroom, hoping that any more shots would go in the opposite direction. Our next stop was in Querétaro, a state not far from the capital known for its university rather than its cartel violence. Several weeks in the lab sawing and crushing rocks, separating the magnetic from the nonmagnetic minerals, and then isolating the coveted zircons using toxic heavy liquid felt like a vacation after the constant undercurrent of tension in Sinaloa. My research collaborator at UNAM let us stay in his guesthouse, and we walked to and from the university lab I was using, sometimes stopping at the bowling alley along the route for an evening drink and a quick frame.

Through all of this, Carlos interviewed for jobs in Australia. We found out he'd gotten a very good one while we were lying in bed at the guesthouse, and it felt like another layer of tension had dissolved. When we returned to Queensland, this time to start our life there in earnest, I would have my newly acquired data to unravel, and Carlos would have a job that would pay well and boost his career up several notches. As far as honeymoons go, ours was unconventional, but that's the life you sign up for when you marry a volcano hunter.

# El Reventador

## THE ERUPTOR

### NOVEMBER 2015, ECUADOR

GRIMACING, I SHOVED the spandex-and-polyester leggings into the last available space in my large expedition pack. Just looking at the things had soured my mood, renewing the internal conflict between excitement at working on an erupting stratovolcano and apprehension at doing the work for a TV film crew for packaging and wholesaling to Discovery Channel audiences as part of their midweek evening programming. After years of turning down TV producers who stumbled across the blog I kept while working at the Hawaiian Volcano Observatory and wanted to create volcano shows that just couldn't work, someone had finally presented me with what I thought would be a scientifically accurate opportunity to show the public what volcano research was really like. I had agreed to the show after Anna, the assistant producer, had plied me with weeks of phone and email conversations, demonstrating she understood the complexity and hazards of volcano science. When I requested an $8,000 Trimble GPS unit, accurate to within a centimeter, to map the volcano's lava flows, she had persuaded the company to loan it out. I had already packed the bright yellow tool, ensuring it was well-padded and would be safe for the long flight to Quito, Ecuador. Since it required a cantaloupe-sized flying saucer of an antenna and a 6-plus-foot-long pole, my luggage for the trip was awkward but manageable. I had everything I needed for two weeks in the caldera of an active volcano just under 4 miles from the official edge of the Amazon rainforest.

My reluctance was because of the leggings, and related issues that had arisen as I coordinated with Anna about the trip. A week prior to my departure, she had emailed pictures of outfits some Discovery higher-ups wanted me to wear. My usual attire for the eight years I'd been doing field research was ideal for the work I did. Caterpillar work boots, sock liners under wool socks from REI, men's cotton khaki cargo pants by Levi's—which I now had to hunt for on the internet, since they were no longer in production—and women's athletic shirts that covered or exposed as much of my arms as the weather dictated. I had learned through repeated trial and error that some items were essential, like those sock liners. I completed the outfit with a cowboy hat, beanie, or bandana, and I hung my thick leather work gloves from a clip on my belt. Every piece of clothing was functional and allowed me to do my work safely.

The Discovery executives on this show must have had more experience clothing people for studio shoots, because the outfits they initially suggested would have made me look like I was starring in a jungle remake of *Little House on the Prairie* or auditioning for a theatrical production of a Charles Dickens novel. They wanted a round broad-brimmed hat, more appropriate for sunning on a beach than tromping through heavy vegetation, a white long-sleeved button-up shirt with oddly ruffled cuffs and a collar I just found confusing, form-fitting women's hiking pants made for being lightweight rather than durable, fashionable women's hiking boots, and the accursed leggings.

When I opened the email, my eyes widened. I shoved my computer in front of Carlos.

"Look at this *shit* they want me to wear! Leggings! On a damn volcano!" I crossed my arms and waited for the outrage I was sure he would share.

Instead, he laughed. That extinguished enough of my indignation that I was able to formulate a rational response to Anna, since I realized it couldn't have been the respectful, diligent assistant producer's idea to suggest I wear clothing that was good at neither form nor function. Synthetic fibers will melt if exposed to high heat, and I knew Anna wasn't trying to get me killed. We negotiated back and forth for a few days, until I had prevailed using my arguments about safe performance on volcanoes. Well, prevailed in every matter except for the damned leggings. The night before I was due to fly out Anna emailed one final time, imploring me to bring the leggings—just in case, for use only around camp. As soon as they were stuffed deep in the bowels of my luggage, they slipped my mind as anticipation for the upcoming adventure took hold.

▲

The vision of the Discovery Channel executives slowly revealed itself as the team assembled in the small tourist lodge at the base of the volcano's caldera. The journey from Quito had taken several hours for our rented vans, climbing the twisted two-lane road between towering mountains, the majority of which were sleeping volcanoes. I sat near Jeff, a volcano geophysicist from Boise State University, whose work on volcano acoustics was familiar to me. He was in his early forties, tall, and in excellent shape from his hobby of competing in Iron Man triathlons. Like many scientists I enjoyed working with, Jeff was sharp and demonstrated both his love of science and a sense of humor right away.

Jeff had worked at Reventador more than twenty times before, alongside colleagues from the Geophysical Institute in Quito. The Ecuadorian scientists do a tremendous job on the frontlines of monitoring the country's multiple hazardous volcanoes. Visiting scientists, like Jeff, contributed to their efforts, but the Ecuadorians were the unquestioned leaders in science, monitoring, and fieldwork in Ecuador. As we traveled, we dropped our voices to discuss our shared misgivings about the show's reluctance to incorporate local scientists in the on-camera work, and agreed to advocate for their inclusion when we could, even if it was most likely a losing battle. Programs with genuine science are few and far between, and we knew we could only become effective advocates for truly representative science if we proved ourselves to the networks.

We were sitting close to Shaun, a former Navy Seal who had served in Afghanistan and Iraq and had most recently been doing contract work in Afghanistan. He had a wild, dark beard that he claimed helped him interact with the locals where he worked, and he came across as humble, funny, and instantly likeable. We alternated between staring at the scenery and swapping stories about work we'd done before. Shaun, Jeff, and I were the "on-camera talent" for this show, which we'd been told was called *Trailblazers*.

The premise of the program was that exciting new discoveries are still waiting to be made in the name of science, and the channel would follow three teams on scientific expeditions around the world. Team Volcano, as we dubbed ourselves, was headed for El Reventador in Ecuador. One of the world's most active volcanoes in 2015, Reventador was erupting about every half an hour. These were small eruptions, yet still significant enough to send a plume of ash and gas a few miles into the air, and most produced short pyroclastic flows. A few years before our shoot,

Reventador had erupted in a larger fashion, sending enough ash to Quito to shut down the airport and government for a few days. Aside from the small guest lodge on the lowest flanks of the volcano's outer edifice, no villages were close enough to the volcano to be in danger from any lava flows it might produce. Reventador, like many volcanoes, carried its greatest threat in the ash it could belch out. Pulverized rock fragments are heavy, so volcanic ash has the potential to collapse roofs, smother crops, contaminate water supplies, and shred lungs if inhaled. I was practically vibrating with energy, keyed up and ready to lay eyes on this very active volcano that was so different from the ones I knew so well in Hawai'i.

One of the other teams Discovery had built for *Trailblazers* was in Bolivia, hunting the elusive black caiman, a crocodile-like creature that held unique, poorly understood genetic attributes. The final team was in remote Papua New Guinea, searching for new species of plants and bats. When Anna had explained the series to me, she said that Discovery was interested in showing scientists who were blazing new trails with their work. She mentioned that they would be teaming us with special forces fighters to make sure we could do our work safely. I thought that was superfluous, since field scientists already work in remote, dangerous places without such help, and we weren't going to be in an armed conflict zone. Meeting Shaun and Jeff in person was a relief, since they were both good-humored and professional. Shaun expressed his interest in learning about volcanoes, so Jeff and I had given him several mini-lectures by the time we arrived at the lodge.

The crew for the shoot consisted of about two dozen people. Most were from the production company, but several local folks had also been hired to work with us. The shoot's doctor was Ecuadorian, as were the young men who had spent the previous week hacking through the jungle with machetes, readying areas in the caldera that the advance team had identified as likely shooting locations. When I emerged from the van into the lodge parking lot, I collided with a heavy, humid rainforest scent mixed with crisp mountain air. It was as if someone had taken the clear atmosphere of the Rocky Mountains and stirred it up with the distinctive musk of the Everglades. I filled my lungs a few times and stretched, happy to be out of the van, and then turned to help the crew with the chaos of unloading the gear and moving it into the basement of the lodge for organizing.

We were given time to settle into our rooms, and then the director, Tom, called us all to meet in the lodge's dining room. Everyone was introduced, and Tom gave us a rough outline of how the shoot would work. That night we'd have time to repack our gear after the crew passed out jungle boots, waterproof duffel bags, and other items we might find necessary in this Amazon-adjacent volcano. In the morning, our first shoot would happen. It would be me, Jeff, and Shaun in the bed of a small pickup truck, and the main camera operator, Danny, would film us as we drove along a road with a view of Reventador. The second camera operator, Jim, the sound guy, Pete, Anna, and Tom would follow us with a "chase car," getting footage from different angles and radioing instructions to our driver inside the truck's cab. It sounded straightforward enough.

After they got what they needed on the drive, we'd head to a field several miles from the lodge to await our ride into the caldera. Instead of making us hike several miles through hip-deep jungle muck carrying our gear just to get to our camp, Tom and Anna had rented a helicopter from the Ecuadorian military. I broke into a grin when I heard, and Jeff asked if I liked helicopters. I told him I loved them, and he explained that he hated them. He was a physicist, after all, and helicopters are in a constant state of controlled falling. It's different than fixed-wing aircraft flight, so I could understand why he wasn't excited about taking a military helicopter into an actively erupting volcano. I wondered how I would be able to sleep.

Once Tom finished detailing the filming logistics for the next morning, he introduced our expert safety crew. It consisted of two former British Royal Marines Commandos, Aldo and Daz. Aldo was Scottish and had been a sniper when he was with the Royal Marines. Daz had served with Aldo, and they now did work for film and TV crews around the world. Both were in excellent shape, and I was impressed when Aldo said his company had overseen the safety rigging for the stunts on *Avengers: Age of Ultron*. It seemed that Discovery was taking safety on the volcano seriously. Aldo told us that he, Daz, Tom, and a crew of

Ecuadorians had been scoping out the caldera for the last few weeks. They'd talked with Tom about shots we'd need to get and identified and inspected places where we could get them. They already had a base camp set up for us, so once we landed in the caldera we could focus on the filming.

▲

The helicopter beat its way through the thick jungle air, the wind rushing through the open side door snatching the breath from our lips and making communication nearly impossible. I pressed my face to the round glass window next to me, where I saw massive waterfalls breaking up the overwhelming green of the forest canopy, the water tumbling endlessly far below. I looked up, across the horizon to where giant volcanoes sat in stately quiescence, the suspended animation of an entity that isn't alive yet still births itself over and over between periods of slumber so deep they lull humans and animals alike into forgetting their true power. We crossed the rim of Reventador's horseshoe-shaped caldera, one side missing due to a catastrophic eruption and subsequent collapse of the rock that was once there. The resulting landslide must have choked the river valley at the foot of the volcano, obliterating everything in its path. In the thousands of years since, the crags left behind with the collapse had been blanketed over by the lush, consuming vegetation of the Amazon basin. The abundance of rain guaranteed that no matter the destruction Reventador wrought, the wreckage would soon be reclaimed by the jungle.

As we skimmed the rim and entered the old caldera, I strained to see the cone of the volcano, the new mountain that had been building since a debris avalanche decimated the volcano's original structure well before humans ever traversed its flanks. New lavas were thrust to the surface and laid down, layer after layer of eruptive material, like a dry version of Oregon's Crater Lake with its resurgent dome poking out of the deep blue waters. The layering process is what earned this type of volcano the "stratovolcano" designation. If you looked at a cross section of the volcano, the different eruptions would be visible, the volcano's history written in stratified rock until the next cataclysmic eruption shattered the tableau. The job of volcanologists is to piece together the histories contained by the rocks. It is only through understanding what volcanoes have done in the past that we can have any idea of what they might do in the future. Understanding how volcanoes can erupt saves lives.

Reventador's cone was cloaked in clouds that obscured the higher reaches of the caldera. The helicopter made a few passes, hopeful the clouds might dissipate enough for a good bird's-eye view of the dome. When the clouds proved stubborn, the pilots took us toward a flat expanse near the open end of the caldera. The helicopter set down, and we piled out. Tom had us grab our personal bags and a few pieces of equipment in silver cases, and instructed Jeff, Shaun, and me to wait about 50 feet from the helicopter. We watched as the crew hurried on and off the helicopter, moving all the supplies we'd need for nearly thirty people to survive two weeks inside the caldera of a remote Ecuadorian volcano. Tom directed that once the helicopter was ready to leave, he wanted Jeff and me to crouch and have Shaun act like he was helping shield our heads from the wash of the helicopter's blades as it took off. That struck me as odd, since I'd never needed anyone to make sure I stayed crouched when a helicopter took off nearby. Jeff and I exchanged a glance but knelt as instructed.

After the helicopter ascended and its rotor noise disappeared, Shaun was told to lead us toward a moss-covered clearing a short distance away. We shouldered our packs and started jogging, with Danny and Jim darting around us with heavy cameras on their shoulders. Tom exhorted Shaun to tell us to hurry up, and as Shaun turned back to do so he tripped and sprawled onto the mossy caldera floor. Danny, adhering to the reality TV filming policy of keeping things rolling continuously, shoved the camera's huge lens down toward Shaun to catch his reaction. Shaun, already laughing at himself, realized it was all being caught on camera.

"Fuck fuck fuck fuck fuck fuck fuck! Now you can't use this!" he exclaimed through peals of laughter, knowing inserting the obscenities would render the footage useless. Jeff and I were laughing too, and so was everyone who'd seen it. So much for the serious, unrelenting danger of working on an active volcano.

Tom made us go back and reshoot the scene, this time sans Shaun's face-plant. After, Tom filmed Shaun heading off into the trees to supposedly go cut sticks for us to make a shelter with. Then the crew filmed us setting up our camp—using sticks the Ecuadorian crew had macheted to the correct size in advance of our arrival. Jeff and I rolled our eyes, beginning to understand how much of this expedition would likely be influenced by TV magic.

After camp was set up, Jeff and I decided to walk around the area. One of his students had an infrasound array deployed around the caldera and Jeff was planning to find time to swap out the array's batteries for better ones he'd brought along. I wanted Jeff to give me the lay of the land, since it's always smart to employ all knowledge of the area when working around active eruptions. Shaun, Tom, and the crew followed us to the edge of a cliff that dropped over 30 feet down to a watershed channel. There was so much rainfall this close to the Amazon that even lavas erupted within the last decade or so had already been cut through by the water. That's lightning fast on the geologic timescale. I was intrigued, and Jeff and I were deep in conversation close to the edge of the cliff when we felt a deep rumble telegraph through the earth and into our feet. We snapped around to look where Reventador's cone lurked behind the clouds and saw that the view was now unobstructed. Just a split second after we felt the rumble in our legs, we saw a bloom of dark ash shoot straight up out of Reventador's narrow throat and into the sky. The guttural boom of the eruption reached us just after we saw the plume, resonating in our chests before the sound wave reached our ears.

Jeff and I yelled in delight, and we both jumped, fists pumping in excitement at the raw power of the eruption. I saw the light gray of the pyroclastic flows leave the summit, speeding down the steep flanks of the cone. The material in those flows was too heavy to be carried upward by the blast, and so it fell out and tumbled down the mountain in clouds of rock and ash heated by the bowels of Reventador. Pyroclastic flows from the 1980 eruption of Mount St. Helens traveled at speeds close to 500 miles per hour. The ones we watched streaming down Reventador were babies by comparison, but I didn't care.

"Yes!" I bellowed as loudly as I could manage.

I shot a look at Shaun, Tom, and the rest of the crew. Every single person was gaping at the eruption column, eyes open in amazement.

"Are we . . . are we *safe*?" I heard a voice ask quietly, with a note of genuine concern.

"Yeah, definitely! If you see us running, that's when you run. Until then, you're fine." I was grinning, too jazzed by the eruption to play the role of serious scientist.

Jeff joined in, "And if we're running, it's already too late. Might as well kiss your ass goodbye."

We turned back to the volcano, oblivious again to the crew while we stared at the charcoal gray plume, now stretching at

least 4 ½ miles high and still growing. From behind us, we heard
Tom ask a question.

"Can you do that again? For the cameras, this time?"

▲

The setup of base camp required us to adjust to the divide
between the on-camera folks and the crew. The crew was full of
TV professionals who seemed comfortable with the arrangements
from the start, but it was awkward for Jeff, Shaun, and I to hear
our sleeping and restroom tents referred to as "hero camp" and
the crew's sleeping and restroom tents simply as "crew camp." I
knew it was just shorthand, but it was a little disconcerting. On
other expeditions I'd been part of, minimal separation existed
between the scientists and the cooks, wranglers, or anyone else.
We had different skill sets, but we were all vital for the successful
outcome of the work. The TV folks were so used to working with
on-camera "talent" that their willingness to accommodate us was
routine for them, but somewhat disorienting for people who are
used to handling everything on our own.

Jeff and I were anxiously awaiting our opportunity to crack
into the science missions we were promised, but the director and
film team had other priorities. The notion of unscripted TV programs operating without scripts was false, and Tom had a list of
scenes he wanted to capture in order to weave a coherent nar-
rative of our time on Reventador. He and Aldo asked to speak

with me privately, over by the cliff where we'd witnessed our first eruption. The volcano erupted every thirty to sixty minutes, so it had become routine to be mid-conversation when someone would point excitedly at a new plume chugging upward from the volcano's summit. Everyone turned and stared, and then resumed talking after the sound of the eruption had faded away. It was a reminder of how quickly humans can adapt to almost anything, given consistent exposure.

When I reached the edge of the cliff with Aldo and Tom, a light rain was starting. They described what they wanted to film, and the more I heard the less I wanted to do it. Tom explained the script had Shaun as the trailblazer whose job it was to assist hapless lab-bound scientists doing research in dangerous situations. He wanted to stage a stunt where Shaun could save me from danger. Hearing that made my stomach knot, since I hadn't signed up to shoot *Avengers 3: Volcano War*. I had never, ever been the damsel-in-distress type, and I prided myself on my good judgment and level-headedness during all types of crises. The anger started to simmer in my gut as they detailed their plan, and it was all I could do to force myself to stay quiet and listen.

Tom wanted me to be walking near the edge of the cliff, looking at the geology, "slip" as the cliff edge gave way, and start sliding over the edge. The plan was to have me grab onto a tree root and hang there until Shaun could throw a hastily fastened lasso made from climbing rope over my shoulders and pull me up to safety. I surprised myself by being thoroughly polite with my response.

I told them I had expertise in geologic hazards in general, not just volcanoes, and as such I would never walk on the edge of a cliff made of poorly consolidated, weakly compacted sediments and debris, and especially not while it was raining, and when the drop was more than three stories. It seemed that they initially thought I was afraid to do what they asked, so Aldo went into detail about the rock-climbing safety harness they would have me wear, and how they would belay the rope, and how I wouldn't really be hanging on to just the tree root, and how it would only *look* like I was in danger.

They completely underestimated me. I wasn't afraid of hanging off a 35-foot-high cliff. I was afraid of being treated as a helpless woman, a nerdy lab rat in need of saving whenever I set foot outdoors. I was afraid that scientists would once again be portrayed incorrectly in popular media, and that it would be my own damn fault for unwittingly agreeing to do a show where I was asked to do stunts instead of science. I told them this. They listened, and then grabbed the cameraman Danny to help convince me. Danny was the main camera operator for TV star Bear Grylls, and he explained how this was all common and accepted on Discovery, and how they would make it look like a total accident, something that could have happened to anyone. All three of them worked on me, saying the executive producers back in the United States were expecting this from them . . . from me. I was pissed.

I told them I would need a minute and went in search of Jeff and Shaun. I told them what Tom wanted me to do, and how the show wasn't treating *all* of us as trailblazers as promised, but instead subordinating the scientists to the special forces. Then I started in on how I should have known better than to agree to doing the show right around the time they asked me to wear leggings, and the look of amazement on their faces stopped me mid-rant. Rather than brushing aside my anger and concern as silly or baseless, they asked me to clarify. Jeff and Shaun were incredulous as they immediately understood the impracticality of wearing leggings around active volcanoes. They agreed it was exploitative, and then they both said that if I had to wear leggings, they would wear leggings, too. Jeff assured me they could make this happen, since he had a few pairs of men's leggings he used under his running shorts. My tension evaporated as the three of us pledged to do our best to keep the show's bullshit content to a minimum. Shaun walked back with me to talk with Tom and the crew.

I agreed to do the stunt if they could make it seem like a total accident, and if there was zero science tied to it. They promised they would have something similar happen to Jeff later to appease my concern at being stereotyped as a helpless female. I didn't like it, but I really didn't want to anger the people I was supposed to work with for the next two weeks on our first full day of shooting. Shaun revealed even more of his good nature as we prepared, making sure I was as ok as possible with every step of the proceedings. I rappelled over the side of the cliff and had to yell out to make it seem like I had fallen. I dangled off the side of the cliff, using the harness to support my weight, bouncing absently off the cliff face while I waited for them to set up Shaun's part of the shot. My anger simmered within, set to a low boil, while I tried to enjoy the up-close view of the cliffside, one I would never have had without the aid of the harness. I started identifying chunks of different lavas while I waited, something to keep my mind on the science and away from my outrage at performing a stunt when I thought I'd been hired as a volcanologist.

We finished the stunt to the satisfaction of Tom and the camera crew, and Shaun and I walked back over to hero camp where Jeff was calibrating his seismic equipment. I pulled out the fancy $8,000 GPS unit and started preparing it for when I hoped we'd get to do some real work. Shaun asked us both lots of questions about our equipment, research, and other volcanoes we'd studied. He told us about his kids, and then Jeff and I talked about our families. By the time they needed us for the next scene, the three of us were a unified front dedicated to maintaining our integrity despite the whims of reality television.

▲

The weather at our camp was mercurial most days, with violent downpours interspersed with hours of clear blue. One evening early in the shoot, we were filming some of the science objectives when we noticed angry thunderheads overtaking the northern wall of the caldera. Jeff, Shaun, and I started discussing the dangers of lighting, and someone joked about making a light-ning rod. Tom jumped on the idea and we rushed to assemble a makeshift lightning rod out of a few lengths of metal borrowed from other equipment. By the time we planted it in the ground 200 feet or so from our tents, real lightning and thunder were adding their flashes and booms to the background. Our tents were on a flat area that was partially encircled by a dry stream channel, at only a few feet above the channel bottom. The rain was sheeting down, and we heard a low rumble that was quieter and less ominous than one of Reventador's eruptions but none-theless concerning. We ran to the edge of the channel and found a torrent of water where only smooth chunks of lava had been minutes before. The waters were powerful and deep enough that anyone standing in the channel would have been thrown off their feet and swept downstream. The crew kept the cameras on us as we exclaimed about the power of the flash flood, capturing some of the real-life drama inherent in field research.

Tom seemed pleased with the impromptu footage. As we hur-ried back to camp in the rain, Jeff and I discussed our hypothesis that the crew would get what they needed in terms of dramatic footage if they just allowed us to do our real work. The challenge of how to bring genuine field science to a public used to consum-ing an Indiana Jones and reality TV version of science seemed enormous. The volcano rumbled in the background; an erup-tion hidden by the storm but happening, nonetheless. Before we entered our separate tents for the night, Aldo informed us that the locals had reported the rain would last for the next three days. We'd need to incorporate the rain in the shooting while simultaneously trying to work around it for some scenes.

The flap of my tent opened to a view of Reventador's cone. Each time I emerged I felt a strong pull to hike toward the volcano. We were a little more than a mile from the summit, although still within the caldera itself. The caldera's horseshoe shape opened to the east, toward the Amazon basin. Our camp was about 6,900 feet above sea level. Reventador's summit was nearly 11,700 feet high. The cone was steep, looming over the caldera with not just the promise, but the regular delivery of violence. The pyroclastic deposits that blanketed its slopes were unstable, and any attempt to summit on foot would be impossible due to the debris littering its sides and the more obvious frequent eruptions. We were too far away to see the lava bombs the volcano was throwing out during the blasts, but I daydreamed about their size and quantity.

El Reventador is a very different sort of volcano from those in Hawai'i. It has more in common with the volcanoes of the Pacific Northwest of the United States, which are the classic cone-shaped type drawn by children (and, admittedly, most adults) the world over. These volcanoes produce lavas that are richer in silica, which makes them stickier and more viscous than the runny, silica-poor volcanoes in Hawai'i. Sticky lavas mean more-explosive eruptions, and they're classified broadly into andesite, dacite, and rhyolite. Reventador produced mostly andesites, which are the runniest of the sticky lavas. Still, it was extremely unlikely that we would find any ropey pahoehoe type flows that are so common with basaltic volcanoes. I knew if we saw intact lavas, they would most likely be crumbly a'a flows or the even more rare block lava flows.

Every free moment I had was spent exploring the numerous stream channels around our campsite. Most of the caldera floor was covered in thick mosses and other water-loving plants. Finding actual bare spots was hard, thanks to our proximity to the Amazon. The stream channels flooded often enough that clean rocks were still visible there, and I took every opportunity to crack chunks of lava with my rock hammer. I held my magnifying hand lens up until my eyelashes skimmed the glass, a milky crystal of plagioclase feldspar or a glinting black pyroxene winking back at me. It wasn't as exciting as working on fresh lava flows, but it sufficed while we were checking off boxes on Tom's list of essential scenes. Jeff and I were both anxious to get closer to the real action, him from experience with this volcano, and me from knowledge of working on others. It felt like we were first-string players sitting on the bench during a championship game, watching and appreciating the spectacle but not yet part of the action.

▲

The shooting script finally called for us to move closer to the cone. The plan was to spend the next few days shooting scenes on a lava flow that had erupted three years before. After over an hour of hard hiking, we reached the flow's edges. This was a block lava flow, which I equated to the closest thing to science fiction in the world of lava. The lava's viscosity was very high and it advanced down the sides of the volcano via a combination of factors, including gravity pulling on the sticky molten material, which broke into chunks the size of mini-refrigerators and solidified. The top of the flow had some jagged, aʻa-like features, but most of the pile was cloaked in a 3-inch-thick layer of moss. It looked as though a giant child had a tantrum and dropped a bucketful of dark green Lego bricks along the north wall of the caldera. At its widest point, the block lava flow measured more than 500 feet across and nearly 100 feet thick. The area of the flow closest to its source, the vent at Reventador's summit, had been obscured by the volcano's ongoing eruptions. It became identifiable near the point where the steepness of the current cone met the older caldera floor and tracked just shy of a mile from beginning to end. It was a beast.

The individual blocks of the flow were jumbled, with gaps the size of tree trunks next to spaces the width of a coin and everything in between. Walking on the flow was an exercise in patience. The moss covered every upward-facing surface of the rocks, like a Chia Pet from hell. My trusty work boots were barely better than going barefoot. The tread did nothing to grip the moss, and every step required precise balance and careful positioning. One wrong step would elicit a face-first dive onto the pile of lava, or even a snapped ankle or leg bone. I did not relish the idea of being evacuated from the volcano via helicopter, so I worked hard to maintain equilibrium between my uncertain feet and my weighty field pack. Watching the camera crew in the drizzling rain, scrambling over the treacherous lava flow to get ahead of us, inspired deep respect for their work. Not one crew member complained despite their heavy, awkward gear and the difficult terrain. Seeing how hard they were working to do their jobs made me renew my commitment to delivering the best TV science I could . . . until it was time for another stunt.

Fortunately, this one centered around Jeff doing something ridiculous instead of me. He and I were talking about something geologists sometimes do in the field, even though we really shouldn't. It's called trundling rocks, and it involves pushing large boulders down hills for no scientific purpose whatsoever. It's just fun. We were climbing a knife-edge ridge on one side of the flow where it had channelized. The ridge was a levee that formed when the flow reached its highest point and then receded a bit, and there was a matching one across the now cool and solid river of lava blocks. The drop-off from the ridge was steep, and jagged, misshapen lava boulders measuring about 3 feet across stood sentinel along the ridgeline. Jeff, Shaun, and I had stopped for a quick snack break while the camera crew was up the ridge about 100 feet ahead of us. Jeff asked if I'd ever trundled rocks, I said yes, and he used his legs to dislodge a boulder. It slowly tipped over and into the channel, at first moving in slow motion and then picking up speed as it crashed down to meet its fellow rocks below. All three of us whooped, sharing the primal excitement that comes from breaking stuff. Rock successfully trundled, we resumed hiking and caught up to the waiting crew. Tom wore a mischievous look, which I had come to associate with him requesting something of us we wouldn't want to do.

His plan was to have Shaun and I exploring down in the middle of the channelized flow, some 30 feet beneath the ridge. Jeff would be up on the ridge and "accidentally" dislodge a boulder. Said boulder would come crashing down, narrowly missing bludgeoning us on its descent. Jeff would then have to claim responsibility, while Shaun admonished him for his carelessness. In real life, I knew Jeff would never be careless when people were working below him. Falling rocks are a regular hazard for geologists, but for the glory of TV he grudgingly agreed to the role. We were both aggravated by how little actual science was included in what the powers-that-be wanted from the shoot, but our only other choice would be to walk out on the production. That would mean both an actual walk out of the caldera and abandoning the fantastic crew, who were doing the best they could with what they were told they needed to accomplish. Jeff and I spent some time talking about how we would make an engaging science show if we were ever given creative license, and we agreed that if we cut and run on Discovery, we'd kill any future chances at improving the genre.

Since this scene was unplanned, Aldo and Daz were quickly brought in to determine which boulder they deemed safest and where they wanted Shaun and me to stand to look convincingly in danger yet still safely away from the rock's flight down. The boulder they chose was nearly 3 ½ feet high, and almost as wide across. It was firmly anchored to the flow, even after Aldo sat down and pushed against it with his back. While Aldo and Danny strategized about the safest way to get the shot, I offered Daz my rock hammer as a method of chipping away at the boulder's base until it was narrow enough to break when Jeff "accidentally" pushed it. I climbed down the side of the channel, catching up with Shaun, who was already below. While the crew sorted out logistics, I was able to look around.

This was my first time inside a channelized block lava flow. The levee walls were so high I couldn't see Reventador's cone at all. The typical flow structures visible on the more fluid types of lava were missing, and it was disconcerting to imagine the sound the blocks must have made as they broke apart from the main flow. A'a flows had a distinct crunching sound, like large teeth chomping on broken glass, and sometimes smaller a'a pieces made a lighter, tinkling noise. I closed my eyes, trying to imagine the noise made by solid pieces of rock, the size of filing cabinets, falling over each other. Just as I was getting a sense of it, Tom yelled down that it was time to shoot.

Danny's stroke of genius was to take one of the smaller, secondary cameras and put it on the hillside, just to the side of the predicted path of the boulder. He hoped it would catch the rock speeding by and create some real visual drama. Jim was stationed down in the channel just behind me and Shaun to capture our reaction. Aldo and Daz were stationed behind the boulder, out of sight of Jim's camera and the one positioned on the slope. Jeff's head was just visible above the ridgeline, and on Tom's count Aldo and Daz pushed the rock.

"Rock!" Jeff's shout bounced off the far wall of the channel.

Everyone watched the boulder teeter for a moment before gravity began pulling it down. It tumbled end-over-end down the slope, tearing up chunks of grass and mossy muck, bouncing swiftly downhill. It shot past us, a good 20 feet ahead of where Shaun and I stood, before clattering onto the main channel rocks and splitting in two. Jim rushed up to me, zooming in on my face. I had the sudden revelation that if they wanted this scientist to act, then act I would. This would be my Academy Award moment—or at least one worthy of a Daytime Emmy.

"Oh my God! Shit!" I exclaimed, my forehead creased in approximation of the consternation and momentary fright from having actual rock falls nearly decapitate me earlier in my career.

"Watch out!" Shaun warned from a few steps behind, his voice conveying a protective urgency I guessed came from genuine experience being concerned for his fellow Seals in combat.

Jim whipped the camera from me to Shaun then upslope to Jeff, who peered over the edge of the levee, looking guilty. Tom called cut, and Danny bounded down the slope to check on his camera. It had gone flying during the take, and it looked like the rock had struck it during the descent. He retrieved the camera, and it was in one piece except for the long foam-covered microphone. The rock had indeed hit the mic during the fall, tearing it away from the camera body. Instead of Danny having to account for a destroyed $5,000 camera, he'd only have to justify killing a $300 mic. He told us he'd broken far better cameras to get dramatic shots before. He pulled up the footage on the camera's viewscreen, and we huddled around to watch.

Even Jeff and I, jaded as we were after several days of seeing our dreams of real volcanology on screen destroyed, had to admit that it was excellent footage. The camera captured the rock barreling toward it, and the whiplash of its tumble after the contact. Danny was thrilled, Tom seemed satisfied, and Jeff and I were relieved the next scene on the schedule was about actual science. We were making interesting TV, but it was still far closer to Indiana Jones than anything approaching reality.

▲

One piece of equipment carried in by the helicopter had remained a mystery to me, Shaun, and Jeff. It was a hard, rectangular, black plastic case just over 4 feet long, and about a foot high. We hadn't seen inside it, and it had been ignored amid the clutter of base camp. So far we'd filmed a few stunt scenes, a hike scene along the caldera rim that required simulated machete work—simulated only thanks to the Ecuadorian team who had cleared the path in advance—and installed a working thermal

camera on the northern rim of the caldera that would be used
by the Instituto Geofísico in Quito. Scientists at Ecuador's Geophysical Institute monitor the country's active volcanoes using
networks of seismometers, normal and thermal cameras, gas
sensors, tiltmeters, and more. In return for scientists at the institute providing us with information about Reventador's current
known activity, the production team had agreed to have us install
the new thermal camera during our shoot. Jeff also had time to
check and upgrade his graduate student's infrasound array, and
the crew filmed us working on that. We were pleased that some
real science was making it onto film, even if we had no guaran-
tees it would survive the editing booth.

The first time we thought to pay attention to the mystery case
was when it appeared at the foot of a landslide we'd hiked to near
the base of the north caldera wall. The slide was fresh, likely trig-
gered by recent rains. About 100 feet of naked vertical rock stood
out starkly against the heavy vegetation of the caldera walls on
either side of the new scar. The slide's toe, or bottom end, measured about 75 feet across. The debris flow at the bottom was
made of tan mud and dark rock and choked with still-green plant
matter. After making that sharp ascent, the landslide cut back,
away from sight. By backing away from the toe of the slide, I
caught sight of the shape of the caldera wall beyond the vertical
face of the slide. A shelf of rock angled back, away from us, and
gradually upward. A half-moon shape traced where the barren,
muddy rock met the undisturbed jungle above. This was the

head of the slide, the area where saturation of the soils caused the weight of the earth and jungle there to detach a chunk of the caldera wall and send it flowing downhill to the caldera floor below. I shuddered. A landslide like that would have come with no warning. It was fortunate the caldera was uninhabited, since landslides near populated areas are often fatal. This one was relatively small, but the steepness of the caldera wall would have made for fast movement of the debris.

When I walked back to the group, Tom, Aldo, and Danny were engrossed in conversation. I noticed a climbing rope dangling over the right side of the shelf, reaching all the way to the ground. Rain started falling in a steady drizzle as I walked up to the group. Without waiting for me to speak, Tom told me the plan was to have Shaun free-climb up the rock face, then set lines for me and Jeff and we would climb up. I asked how this possibly fit into the story, since there was no reason we'd ever climb the caldera wall on a real expedition. We'd already installed the thermal camera on the rim, and we'd filmed the intense hike required to accomplish that. Tom explained when Aldo's advance team had found the landslide just days before we arrived, they'd written scaling the wall into the script as part of our journey to get as close as possible to the active eruption. My science brain was in full revolt. The rain was still coming down, and Jeff and Shaun came over while I was protesting to Tom and Aldo.

I had experience with all sorts of geologic hazards, not just volcanoes. I knew the mechanics of landslides, and I explained we had no indication the slide was stable or finished moving. The rain we were all feeling could reactivate the slide at any moment or dislodge part of the earth that hadn't yet moved. Landslides are extremely dangerous, and these guys wanted us to climb one in the rain for theatrics. Aldo tried reassuring me, and Tom tried coaxing me. Danny cajoled, telling me how badass it would look on camera. None of it swayed me, and Jeff said he trusted my expertise with respect to the potential danger of the slide.

I felt the tension grip my gut, my knowledge of landslide hazards in direct conflict with the crew's desire to get a dramatic shot. The rainfall until then had been light enough that I doubted the soils were saturated enough to reactivate the slide. That prompted me to offer a compromise: if the rain continued for another ten minutes or picked up intensity, we'd call off the scene. Otherwise, I would do it, grudgingly.

Tom agreed, and the crew got into position. It was only then that the contents of the mysterious hard plastic case came into play. Aldo approached Shaun with a large, unwieldy device that looked like a strange gun. I'd never seen anything like it before, and Tom told us it was a grappling hook. The plan was for Shaun to shoot it up and over the rock face, and the hook would carry a rope that Shaun could use as a safety line. They acknowledged that there was no way it would create a secure line, so Aldo scaled the cliff face using the already-fixed lines to the right of the slide to secure Shaun's line out of view of the cameras. The magic of TV at work again, I supposed. The lack of realism was exasperating, and a glance at Jeff suggested his patience was also being tested. We made some sarcastic comments about how we never visited volcanoes without our trusty grappling hooks. I fervently hoped the rain would pick up so we could get out of making the unnecessary, dangerous climb.

It wasn't the actual climb that bothered me. I was a veteran of working in hazardous conditions, and even though I was not a rock climber I did enough work on rocks that I might as well have been part mountain goat. The pointless drama is what bothered me. While I'd been promised the opportunity to show real-life volcano research to the public, I was instead expected to deliver manufactured danger solely for the purpose of audience titillation. I reasoned the inherent danger in researching active volcanoes would be far more interesting to the public than anything a TV executive could design, but I was in no position to bring a program like that to life. I marinated in my angry thoughts until Shaun positioned the grappling hook pointing upward and fired. With a *thunk* and a *whoosh*, the hook flew upward, cleared the cliff face, and disappeared. A minute ticked by and then we heard Aldo's heavy Scottish burr over the radio saying the line was ready for Shaun.

The former Navy Seal began ascending the near-vertical rock face. His dark green shirt quickly became streaked with rain, and water beaded in his thick beard as he moved upward. He had a length of climbing rope wound around his shoulder, and we all watched as he made his way up, free climbing the slick rock. He wasn't relying on the safety line at Tom's request, which heightened the drama. When we first met Shaun, he told us he'd recently injured his shoulder in Afghanistan. He hadn't had time to get treatment for it before the shoot, and now he was climbing the site of a recent landslide while carrying about 40 pounds of rope through just enough rain to ratchet up the difficulty. If he lost his grip, he'd have to grab the safety to arrest his fall. Still, he climbed steadily, projecting an impressive confidence.

I exhaled when he crested the top of the exposed rock face and waved back down at us. The experienced climbers down on the ground checked the harnesses Jeff and I were wearing, and I looked at the time. We were closing on the deadline of ten minutes I'd set and as if on cue, the rain petered out. I grumbled, but Tom would get his dangerous climbing scene after all. I just hoped I'd been right about the soils farther up the slide being stable enough to stay in place while we climbed. My job was straightforward—simply climb to the top of the slide. The real challenge was for Danny, who would have the 30-pound camera on his shoulder while being hauled up the rock face slightly ahead of the ascent Jeff and I were making. Daz and Aldo were using the fixed ropes and Danny's harness as a pulley to move him steadily upward so he could focus on filming me and Jeff. He would occasionally bounce into a rocky overhang, but he was able to focus on us for the whole climb. As I made my way upward, I distracted myself from thoughts of the landslide with imagining how this scene would look on TV, and how silly it seemed to me for the show to go to such lengths concealing the camera crew when the final footage would obviously reflect that someone was there and filming the entire time. I thought I would much rather watch a show that let viewers in on the behind-the-scenes dynamics between the on-camera team and the crew. A few more minutes of exertion, and we reached the top of the vertical cliff.

I didn't breathe easily until we'd climbed past the head of the slide and into the thicket of greenery beyond. The rain resumed in earnest almost as soon as we'd finished the climb, and I was thankful I'd made it through another day of unexpected stunt work. As we hiked back down from the rim on the trail cleared by the Ecuadorian crew, Jeff and I exchanged a few words about how ludicrous the whole endeavor was. We caught up with Shaun and congratulated him on his impressive climb, and he told us his shoulder hadn't appreciated the effort. His attitude was still phenomenal, and between Shaun and Jeff I was supremely confident in the professionalism of my on-camera costars. We were all trying our best to accommodate Tom's demands, and I could tell we were all more than ready for the scheduled day off from filming.

▲

I was trying to use my free afternoon to read when several members of the team decided to stage a group workout in front of our tents in hero camp during a period of sunshine. Aldo and Daz, being ex-special forces, were in superior shape. Shaun couldn't do as much as them with his injured shoulder, but still held his own. Jeff, veteran of Iron Man races, also kept up, and Tom, Danny, and Jim joined in as well. It was obvious from their physiques that Tom and Danny worked out regularly, and Jim was a former high-level basketball player. I found I couldn't concentrate on reading my book about Kim Jong-Il's filmmaking aspirations while seven sweaty men taunted each other through push-ups and burpees 10 feet from my tent. Watching the display of testosterone in front of the backdrop of Reventador was made even more surreal when the volcano erupted again, a sight so routine now that the guys laboring in front of me didn't even bother looking at it. I rolled my eyes at the good-natured needling between the men and extricated myself from the tent. I reasoned it would be more peaceful and more delicious by the cooking tent.

That evening, Shaun told the crew doctor, José, that his feet were bothering him. The rest of us relaxed with adult beverages in hand, talking and watching as José looked at Shaun's feet. We heard José exclaim when Shaun removed his socks, and a few of us went over to look. Shaun had been wearing the British-made jungle boots provided to us for the shoot. None of us had an opportunity to break them in prior to filming, and the soles of Shaun's feet were entirely consumed by whitish blisters. My memory flashed back to the aftermath of my brutal 'Ainapo Trail hike seven years before, and I shuddered. After cutting my own full-foot blisters open with an X-Acto knife, I couldn't put weight on my feet for a solid day afterward. Shaun's feet looked even worse than mine had, and he downplayed the pain I knew he must have been experiencing. José was alarmed at the severity of the blistering and told Shaun he'd prefer if he could take a day or two of rest. With the tight film schedule we had to follow for the rest of the shoot that idea was a nonstarter. José did the best he could to drain and bandage the blisters, and once Shaun had a beer in hand it seemed like all was right in his world. Again, his sheer willpower and unfailingly good attitude impressed me.

During our evening of relaxation, an inebriated Tom sat down next to me to chat. To be fair, most of the crew was intoxicated. Filming in a remote location is taxing, and we had all been pulling twelve-plus-hour days for nearly two weeks. Tom apologized for the stunts he'd made us do, and he explained more about how it was driven by the Discovery Channel executives rather than his own personal choices. He was employed by the production company that had won the contract with Discovery to produce the series. Their main job was to keep Discovery happy, and that meant obliging the whims of the executives, even if they had zero grounding in science. Tom leaned in close to me, conspiratorially.

"Do you know what they said to me?" he asked.

"No," I answered, genuinely curious.

"They said, don't bother coming back if you don't get footage of the scientists running away from the volcano in terror, yelling, 'she's gonna blow!'" he responded, solemn.

I blinked at him, rolling the idiotic image around in my head. I wasn't sure whether to be offended or amused. I decided on amused and threw back my head in laughter at how preposterous it sounded. Tom joined in.

"I knew within two minutes of meeting you there was no way in hell that would happen. It may mean I don't work for them again, but oh well," he trailed off with a shrug and a grin.

"You got that right." I thought back to how excited Jeff and I had been for the first eruption, and the looks of wonder and fear on the crew's faces at witnessing the same event. At the very least, Jeff and I had managed to show the entire crew how real volcanologists approach volcano research. The satisfaction of that knowledge didn't entirely make up for the frustration of doing scripted stunts, but it made me realize how important it is to elevate the role of field scientists in the public eye. How could we change the public's perception of scientists if we aren't ever seen? We would need seats at the table if we ever hoped to change the conversation.

▲

The last few days of filming were devoted to what I knew would be my favorite part of the entire shoot. We would hike for several hours past the huge block lava flow to the back of the caldera, right where the cone met the caldera floor. We would camp up there and spend the next few days installing a tiltmeter and aiming to collect a fresh lava bomb, and we would be within the range of ejected bombs during every eruption that happened whenever we ventured away from camp. We would carry in all our water and food, and we would be far from the relative luxury and convenience of the base camp. Only the core film and safety crew—Tom, Danny, Jim, Pete, and Aldo—would be going with me, Shaun, and Jeff to the high camp. Radio communication would be limited, and we divided up the research equipment, food, and water as best we could. After some discussion, we determined we would be short on water based on how much everyone had the capacity to carry. Shaun volunteered to carry extra for the group and refused to be discouraged despite his ragged feet and compromised shoulder. He ended up with an extra 15 pounds of weight in his pack. When I tried to express my gratitude he shrugged it off, still humble and willing to do whatever it took for the team to succeed.

The hike was long and arduous, our focus consumed by trying to avoid falling on the block lava or slipping on the knife edge of the channel ridge. We were soaked by the time we reached the new camp. Night was rapidly approaching, and the intermittent rain was punishing. We scouted a little way ahead of where we would camp, getting a closer look at the terrain we would have to cover to get to the location where we'd install the tiltmeter the next morning. Jeff insisted we all don helmets before we went forward, and when the clouds cleared to expose Reventador's summit, I understood why. We were just over half a mile from the volcano's summit and looking around I could see evidence of lava bombs on the landscape nearby. We strapped the helmets on and headed up a short incline. Just as we crested the hill, Reventador let loose with another eruption.

That proximity to the explosion took the sensory experience to another level. My chest rattled with the force of the explosion, and the sound took almost no time to reach us. It was so loud that for a few seconds, talking was impossible.

"Rock!" Jeff and I yelled in unison, our experience taking over. We motioned the group to follow us into the shadow of a 6-foot-high boulder, which was the only semi-protected area on the sloping shoulder of the volcano. It wouldn't be much help if lava bombs were coming at us, though. Even our helmets were more of a formality than an actual safety tool.

We counted out loud until we were certain enough time had passed for us to be out of danger from falling bombs. The crew wanted information about this new hazard, so Jeff and I explained the gist of what we were facing. I relished the opportunity to be dealing with real danger instead of stunts for a change. We told them that the car-sized boulder we had taken cover next to was a lava bomb, and I watched the crew's eyes widen (except for Aldo, who didn't seem fazed by anything) when I told them these bombs were ejected from the volcano at 190 miles per hour. Jeff and I poked around until we found some smaller bombs, and we pointed out the impact craters left by bombs the size of grapefruits and basketballs. I went into a bit of detail about how the bombs are shaped by their flight through the air, since the rock is often still molten when it's ejected. I pointed out a few cow pie bombs, which look strikingly like their namesake. I described spindle-shaped bombs, which spiral through the air like footballs as they fall to the ground. Satisfied that the crew appreciated the dangers of our new camp location, we headed back down the small hill to get out of our drenched clothes and set up camp.

That night, as we cleaned up after our dinner of boil-in-bag camping meals, Reventador erupted again against the deep black of the Amazonian night. This close to the summit, we watched, awestruck, as hundreds of superheated, electric-red, incandescent rocks seared their way down the volcano's flanks, bouncing and tumbling vividly against the deep charcoal of Reventador's cone. They blazed a fiery network of zigzagging lines down the steep mountain, like the capillary system of a devil. No one spoke for at least a full minute, the raw power of the volcano commanding silence until the last of the hot rollers had tumbled to a halt, their glow finally faded to blend with the dark of the night.

Our reverie was broken by a quick string of obscenities from Danny. "The one fucking time I'm changing the fucking battery on the fucking camera!" His usually cheerful face was clouded by genuine anger, which I instantly understood. He'd had the lens pointed straight at the volcano's summit, poised to capture the spectacular eruption. Bad timing had ruined his chance, and his artistic side was furious at the missed opportunity. We tried to console him, but I knew it would be hard to catch another eruption with that many glowing rocks streaming down the volcano. I filed the image away in my memory, grateful for the experience and hoping to preserve it forever in the only way I could.

▲

The installation of the tiltmeter went smoothly, and Tom was pleased with the footage he caught. The only frustration was the cloud cover that persisted that close to the summit, leaving us with only short windows of unobstructed views of the cone itself. The final scene left to film on Tom's script was one I found both exciting and concerning. He wanted us to use our portable thermal camera to track lava bombs as they hurtled down from the summit after an eruption, and then have me go into the hazard zone to collect a fresh bomb on camera. I had the equipment I needed to do the job, but I hesitated. While Reventador was a very remote, inaccessible volcano, numerous other ones around the world were both actively erupting and had easy access for the general public. I feared that if I made collecting a hot lava bomb look safe and feasible, I'd be somehow responsible when a member of the public tried to imitate me and died or was injured in the process. I explained my misgivings to Tom, and he was disappointed. I told him I had no problem collecting the bomb, but I really didn't want to appear on film doing it unless we had a way of making the audience understand the real danger of being in a zone where bombs were falling.

After some discussion between me, Shaun, Tom, and Aldo, we formulated a plan. Tom could still get his hot bomb shot, Shaun could look like the hero the network wanted him to be, and I could rest assured I wouldn't be enticing tourists to their deaths. I stood at the top of the debris field, perfectly positioned to use the thermal camera on the area where we knew the bombs were coming to rest. Aldo was hidden around a curve in the debris field, to my right. He was in position with a small camp stove and a few fist-sized chunks of rock. Shaun was below me, where the base of the hill of debris contacted the bottom of Reventador's cone. He was equipped with a helmet, a metal pail, and my rock hammer. The summit view was covered by clouds, so we could only see the base of the cone rather than its full height. If an eruption happened while we were shooting, we would have zero visuals on any incoming lava bombs. On Tom's cue, we sprang into action.

I held the camera up to my eye and called out to Shaun that a hot rock was bouncing down the volcano toward him. He picked his way across the debris as quickly as he could, metal pail banging into rocks as he clambered over and around them. Aldo, having heated a rock with the camp stove, gave a yell before he flung the artificially hot rock out past Shaun. I tracked the rock's trajectory as it arced away from Aldo's hiding spot and landed about 30 feet in front of Shaun, just another gray chunk amid a sea of identical gray chunks. Using the thermal camera, I was able to see it glowing white-red against the cooler blues of the rest of the rocks, so I directed Shaun over to its resting place. He nudged it into the metal pail and hurried back across the debris field until he reached me. By the time he made it back, the rock had cooled. Danny and Jim paused the tape, and it was time for more TV magic. Aldo joined us at the top of the debris hill with the camp stove. He heated the rock again and dropped it back in the pail. Danny and Jim rolled again, and I leaned over and spit on the rock. My saliva sizzled as it hit the rock's surface, and I congratulated Shaun on successfully retrieving the fresh sample I needed.

It wasn't exactly the genuine thing, but I was relieved we'd executed the dramatic highlight of the shoot without injury, and without me setting a dangerous example for anyone watching at home. As we hiked back to down to break the high camp and return to the crew waiting at the base camp, Jeff and I spent time talking about the kind of authentic volcano science show we wished we could make. We'd survived filming on El Reventador with our scientific integrity intact, and I felt an abiding sense of gratitude at the chance to share a small window into my love for volcanoes with audiences around the world.

I resolved that any future TV work I agreed to would be projects where I had a say in what would appear on screen. The experience made me realize I needed to question the science at the heart of any show I was invited to work on, and that speaking frankly with the creative folks in charge of the programs would be essential to do shows that didn't just sensationalize fieldwork, but portrayed the harsh realities scientists around the world reckoned with on a daily basis just to do our jobs. In line with my scientific training, I knew this was only one piece of data about filming science shows. I had seen enough stunning nature programs growing up to be certain that doing volcano and other extreme science shows was not only possible, but a real opportunity to engage the public in the endlessly fascinating, always changing world of scientific field research. Unless scientists became more comfortable working with both journalistic and entertainment media, we would never have the chance to show the world the dangerous, enthralling, essential work we do every day to keep people safe. One rocky experience on El Reventador could serve as the runway to dramatic new ways to tell the stories locked in our planet's violent past, and written anew every day in the language of the Earth itself.

# 9

# The
# Explorers

EARLY 2017, LOS ANGELES AND NEW YORK

I SLIPPED INTO the warm wave of applause, returning the audience's enthusiasm with my smile. The museum curator took the microphone from me and thanked me for speaking. As he gave the crowd information on the snacks waiting in the next room, I glanced at the image projected on the giant screen behind me: I wore an olive green flight suit and helicopter helmet and was sprawled on a bed of ropey pahoehoe lava in the middle of making a "lava angel." If the audience's reaction was to be trusted, the highest-profile speech of my professional career had been a success. The hall lights brightened, and the crowd rose from the rows of folding chairs, murmurs forming a quick crescendo to normal conversation. Carlos appeared by my side with a cup of something pale and bubbly, which he handed me before tapping the rim of his plastic cup against mine. He stepped back half a pace, allowing the initial wave of people to arrive.

After I'd spoken individually with about a quarter of the attendees, my friends Jason and Brin approached with their twin boys, Liam and Watson. The boys were four, and looked to be the youngest audience members for my talk at the Natural History Museum of Los Angeles County. Since the subject was volcanoes, I thought it may have been enough to capture their interest. Apparently, I was right; they were staring up at me with enormous eyes, expressions alternating between shy smiles and open-mouthed gaping. Carlos and Brin started chatting, and Jason kept the boys in front of him while we had a moment. We'd been friends since we met while working in the Arizona State Archives in 2005. I was using my new history undergraduate degree, and he was finishing up his master's degree in public history. Now, in a different state twelve years later, a new career for me and two kids for him, his presence meant a great deal to me. He told me they'd enjoyed the talk and tried to get the awestruck kids to chime in with only marginal success. Then, his expression shifted.

"I wish someone like you was in elected office," he intoned, the gravity of his words apparent. "We need science and rationality in our leadership. I worry about the world my kids will grow up in . . . what kind of future they'll have."

I looked down at Liam and Watson, who were now having the kind of secret twin brother conversation outsiders could never hope to fathom. The potential held in their entire lives, in the lives of every other kid who would now have to grow up in this rapidly changing world, hit me like a physical blow to the solar plexus. Eight days before, our country had inaugurated a man who ran his presidential campaign by attacking and degrading science, sowing misogyny, racism, discrimination, and fear, and who pledged to serve an increasingly narrow portion of our population by viciously attacking all who dared differ from his views. The world Liam and Watson would inherit would be a direct result of the decisions we made now, and over the next decade or two. I swallowed hard.

Would they be able to live where they chose, or would sea level rise or desertification reduce their options? Would they be able to work in whatever field they wanted, or would automation and crushing student loan debt push them into jobs that ate away at their natural inquisitiveness? Would they be able to enjoy national parks and wild lands like I had been fortunate to experience my entire life, or would they be forced to learn about the extinction of the grizzly bear or the rhinoceros from textbooks and museum displays? I wasn't thinking about just those two, but rather the colossal ramifications of the election of a president who cared for nothing but his own image and was now in charge of setting the tone of our national and international discourse for the next four years. A lot of damage could be done in that short time, especially in terms of climate and environmental policy. Our world was racing headlong toward a point of no return in the effort to reckon with anthropogenic climate change.

Scientists had warned of the impacts of uncontrolled release of carbon dioxide and other gases into the atmosphere since the 1890s. A dedicated campaign of disinformation and lies about the human drivers of climate change was conducted by many of the same self-serving marketing spin doctors who hoodwinked the public for years about the link between tobacco use and cancer. They had been working for companies and investors tied to the economic success of fossil fuel exploitation, and the marked shift of one American political party from environmental conservation to something better identified as environmental abuse was proof of their success. I and many of my scientist friends had noticed an erosion of public trust and confidence in both science itself and the scientists who were dedicating their careers to stopping the catastrophe in motion that was human-caused climate change. I understood what Jason was saying, and I told him so. As we turned away from each other in search of hors d'oeuvres, a quiet voice in the back of my brain piped up.

*"Why not you?"*

Before I could examine the thought, two scientist colleagues pulled me into their conversation. They were discussing the importance of having science inform policy decisions and wondered aloud at how much damage the new administration would wreak on the natural world and funding for science. The conversation shifted to the executive order the brand-new president had signed the day before, which amounted to a ban of immigrants from numerous Muslim-majority nations. I gritted my teeth in response, listening as they speculated about what would happen to the Environmental Protection Agency, the Paris Agreement, and NASA. Then one of them turned to me, his face earnest

"Why don't you run for office? You're charismatic, you're young, and you're a scientist."

The other scientist joined in, urging me to run. For what, they didn't specify. Before the conversation could go any further, a museum official came over to thank me for speaking. As we shook hands, the quiet voice in my head spoke up again.

"*Yeah, why* don't *you run for office?*"

▲

A couple months later, Carlos and I were back in Manhattan for the Explorers Club Annual Dinner. During the days of events leading up to the dinner, I floated the idea of a congressional run to some club members. Every person I mentioned it to responded positively, with some excited about my scientific background and the prospect of evidence-based policymaking. Others wanted to know about the representative I was targeting, and when I mentioned his denial of basic climate science they reacted with disbelief. Almost no one knew a science-denying Republican held a congressional seat in Los Angeles County in 2017. Some club members just seemed excited about the idea of a club fellow in Congress. I'm not sure what I expected, but it certainly wasn't such resounding encouragement.

As it had the year before at our first dinner, the evening flew by in a blur of handshakes, stories of far-flung corners of the planet, and stiff drinks. This year I opted for a custom-fit men's tuxedo instead of a dress. My hair was spiked into a mohawk, the long ends trailing down behind me like the horsetail crests on the helmets of Greek soldiers. The dinner was held on Ellis Island, and I was grateful for the tuxedo's warmth on the nighttime ferry rides across the frigid waters of the Upper New York Bay. The shaved sides of my head were cold, which was a new sensation. The mohawk was less than a year old, and this was my first opportunity to test its performance in cold weather.

When the ferry dropped us off back in Manhattan, Carlos and I decided to walk uptown for a while. I relished wearing shoes I could walk in as opposed to the usual misery of high heels. Dresses and their associated footwear were receding into memory like a bad dream. We walked briskly up Broadway, trying to stave off the officers of March in New York City, until we came across a famous sight. I remembered the statue *Charging Bull* from visits to the city when I was much younger, and I patted its sleek side before turning my attention to the new star of the narrow strip between the lanes of traffic.

*Fearless Girl* had been installed opposite *Charging Bull* just eight days before I visited, her 4-foot-high bronze figure defiantly staring down thousands of pounds of rampaging animal.

She was somewhat controversial, since she was commissioned by a global asset management firm—not exactly the sort of organization historically known for encouraging women or standing up to the unfettered forces of capitalism. Still, her sculptor was a woman, and her brave stance and undaunted metal features had inspired countless in-person visitors and millions of internet surfers in a little over a week. When Carlos offered to take a picture of me with her I hesitated a moment, reflecting on what the statue meant to me. And then I realized it wasn't about me. None of it was about me.

I was poised to jump into a race for the United States Congress, away from the scientific career I'd worked so hard for and into the unforgiving glare of a national spotlight. I stood there, in a men's suit and a mohawk, a tattooed lover of punk music and animals, a historian-turned-scientist, the child of two Republican FBI agents, the spouse of a first-generation American—a man whose mother was undocumented until he was six years old—knowing myself to be a fighter and a failure, a dreamer, a writer, an athlete, a friend, and a success by the same means anyone else is: I refused to give up. My life was so rich, so full of fortune and opportunity, and I was so desperately in love with our planet. The thought that others may not have opportunity, the injustice I knew shaded far too many lives, and the looming environmental catastrophe I'd seen written in the hard data were the reasons I was there.

That little girl frozen in bronze, defying the odds, is all of us. We are all capable of greatness, because we are human. We are all born as scientists, testing our ideas about the world around us from the moment we open our eyes, seeking connection with another person. Curiosity is our birthright, our shared human heritage that connects us to every other being on Earth, and to the stars beyond. My purpose now was to communicate the breathless wonder, innumerable challenges, and glorious discovery inherent in human existence to everyone who has that curiosity about the unknown lodged deep in their being. We were long past the time for staying within our comfort zones, and this was my chance to leave mine.

I stepped behind *Fearless Girl* and crossed my arms, staring past the bull and into a challenge bigger than any single one of us. I did it because that is what explorers do, after all. We enter the unknown, seeking to illuminate the darkness and charting a course for all those who wish to find a way.

# Acknowledgments

I would like to express my deep gratitude to Madeleine H. Blais, whose support, encouragement, and direction have been a force of nature in my creative life for more than twenty years. This book exists because she does.

I would also like to thank my Goucher College mentors Suzannah Lessard, Philip Gerard, Jesse J. Holland, and Jacob Levenson, as well as the indefatigable Leslie Rubinkowski. They all made me better, both as a writer and as a person. What more could anyone ask for?

Huge thanks are in order to my agent, Sharon Pelletier, who saw the block of marble that was my manuscript for what its final form could be—this book. Stacee Lawmann and Mike Dempsey at Timber Press were wonderful to work with, never nagging and always nudging me to keep everything on track. The edits made by Mollie Firestone were like a carefully wielded rock hammer, revealing the heart of each story for public consumption.

Finally, thanks to the scientists who opened doors for me—including helicopter doors. Kim, Frank, Ramirez, Mark, Matt, Joe: know that I am at my most sincere when I say you all rock. Officially.